O&B
MATHS BANK
2

CHELTENHAM
PATE'S GRAMMAR SCHOOL FOR GIRLS

Name	Form	Date Lent	Date Returned
Emma Hamilton	3R1	20:11:86	....
....	.....	....	....
....	.....	....	....
....	.....	....	....
....	.....	....	....
....	.....	....	....
....	.....	....	....
....	.....	....	....
....	.....	....	....
....	.....	....	....

O&B
MATHS BANK
2

K. J. Dallison M.A.

J. P. Rigby B.Sc.

Oliver & Boyd

Illustrated by David Brogan and Tom Reid

Oliver & Boyd
Robert Stevenson House
1–3 Baxter's Place
Leith Walk
Edinburgh EH1 3BB

A Division of Longman Group Ltd.

ISBN 0 05 003155 4

First published 1978
Second Impression 1980
Third Impression 1980

Printed in Hong Kong by
Commonwealth Printing Press Ltd

Contents

Contents

Foreword

I first came across the authors' enthusiasm as teachers ten years ago, when they founded our local mathematics association for sixth formers and invited me to become its first president.

All mathematics teachers who use one of the newer types of school syllabus will recognise the usefulness of these books. The introduction of any new syllabus should always be backed up by a wealth of problems, not only for the teacher's benefit, but also to provide side paths along which the pupils can explore and play. This is especially true in the case of the more gifted, and especially in the mixed ability classes of today, in which the more gifted must necessarily be expected to work more on their own.

Learning (that is assimilating and memorising) any syllabus can be a tiresome business, and in mathematics at any level there are always opportunities for indulging in the more challenging and exciting tasks of discovery and creativity as well. These require reflection alone, and some form of guidance. There is no better guidance than well chosen problems, that appeal to the intuition and focus the imagination, and through which the student can recreate his or her own mathematics. Such self-discovery leads to a much deeper understanding, and a confidence in the subject, which the student will never forget, and upon which he or she can build further.

E. C. ZEEMAN
University of Warwick
August 1977

Preface

These books are rather different from the usual mathematics texts in that they contain almost no teaching material. They do, however, contain a wealth of questions, covering the modern and traditional mathematics required by 'O' level and C.S.E syllabuses in modern mathematics.

Maths Bank can be used to supplement any existing course in modern mathematics but it can also be used as a course book in its own right, leaving the teacher free to instruct in his own way.

The questions are designed to cater for a wide range of ability: each section begins with easier questions; harder and deliberately wordy questions are starred.

The authors wish to express their gratitude to Miss P. M. Southern, Mr E. P. Willin and other past and present members of the mathematics staff at Rugby High School for writing questions and supplying answers; to Mr M. E. Wardle, head of the Department of Mathematics at Coventry College of Education for acting as advisor on difficult points; and to Miss D. M. Linsley, former headmistress of Rugby High School, without whose foresight in allowing the school to change to modern mathematics in 1963 these books would never have been written.

K. J. DALLISON
J. P. RIGBY
1978

Similarity and Enlargement

1A

1 In your book draw an irregular figure such as the one shown and using a point *O* outside the figure, enlarge it by a scale factor of 2.

Enlarge the original shape again,

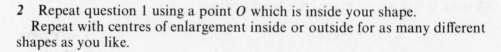

    a) by a scale factor of 3,
    b) by a scale factor of 4,
    c) by a scale factor of $\frac{1}{2}$.

What does *c*) actually do to the original shape?

2 Repeat question 1 using a point *O* which is inside your shape.
  Repeat with centres of enlargement inside or outside for as many different shapes as you like.

3 On graph paper using the same scale on both axes, plot the points $(-1, -1)$ $(2, -1)$ $(2, 2)$ and $(-1, 2)$. Join up to form a square.
Using $(0, 0)$ as the centre of enlargement, enlarge the square by

    a) a scale factor of 2   *b*) a scale factor of $2\frac{1}{2}$ *c*) a scale factor of $\frac{1}{2}$.

Write in the co-ordinates of all the vertices of all the squares.

4 Draw the original square from question 3 and enlarge it using a scale factor of 2 from the centre of enlargement $(1, 1)$. Write down the co-ordinates of the vertices of this square.
  Find the areas of the two squares. How many times larger in area is the second than the first?

5 On graph paper plot the points $(-1, 0)$ $(1, 0)$ and $(-1, 1)$. Join up and make a triangle. Taking $(0, 0)$ as the centre of enlargement, enlarge this triangle

    a) by a scale factor 2,   *b*) by a scale factor 3,   *c*) by a scale factor $\frac{1}{2}$.

Write down the co-ordinates of the vertices.

Write down the areas of all the four triangles which you have drawn.
How does the area of each enlarged triangle compare with the area of the original? How do these area scale factors compare with the scale factors you used for the lengths?

6 On 2 cm gráph paper plot the points A (2, 0) B (1, 1) and C (−1, 1). Join up and make the triangle ABC. Taking the centre of enlargement P as the point (1, 0) enlarge ABC by a scale factor 2, to make triangle A' B' C'.

On your diagram write in the lengths of AB, BC and CA and A' B', B' C' and C' A'. Write in also the lengths of PA, PB, PC, PA', PB', PC'. Mark in your diagram all the pairs of parallel lines which you have drawn.

7 Draw a square of side 2 cm. Using a point P inside the square (not at the intersection of the diagonals) as the centre of enlargement, enlarge the square by a scale factor of 3. Write in the lengths of all the lines joining P to the vertices of the two squares. How long are the sides of the enlarged square? By how many times is its area larger than the original square? Mark all the sets of parallel lines you can find in the diagram.

8 All squares are similar to one another. What other shapes are always similar to one another?

9 $ABCD$ is a square of side 1 cm. $PQRS$ is a square of side 3 cm. If O is the centre of enlargement which maps $ABCD$ on to $PQRS$ and BP is 1 cm, how far is O from A?

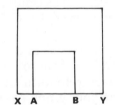

10 The square on XY is of side 4 cm. The square on AB is of side 2 cm. If XA is 0·5 cm, where is the centre of enlargement which maps the small square on to the larger one?

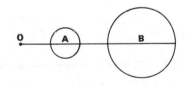

11 A is the centre of a circle of radius 1 cm. B is the centre of another circle of radius 2·5 cm. If O is the centre of enlargement which maps the small circle on to the larger one and $OA = 3$ cm, what is the length AB?

12 In triangle ABC, $AB = 2$ cm and $AC = 3·5$ cm.
O is the centre of enlargement which maps triangle ABC on to triangle LMN. If $OA = 1$ cm and $OL = 1·5$ cm, how long are LN and LM? What is the scale factor of the enlargement?

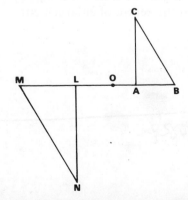

13 Copy the diagram into your book and draw construction lines to find the centre of enlargement which maps the inner rectangle on to the outer one.

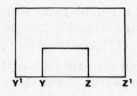

14 Triangle *ABC* is equilateral and of side 1 cm. The other triangle is also equilateral and of side 3 cm. Copy the diagram and mark the centre of the enlargement which maps triangle *ABC* on to the other triangle. Letter its vertices *A′ B′ C′* as appropriate. What is the scale factor of the enlargement?

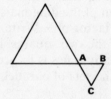

15 On graph paper, draw axes taking values from -12 to $+8$. Plot the points $A(2, 2)$, $B(6, 2)$, $C(6, 4)$, $D(4, 6)$, $E(2, 4)$ and join up to form a pentagon.

Using $(1, 0)$ as centre, enlarge the pentagon, using a scale factor of -2. Label the vertices *A′ B′ C′ D′* and *E′*.

16 On graph paper, draw the *x* and *y* axes, taking values from -8 to $+8$. Plot the points $A(2, 2)$, $B(2, 8)$, $C(6, 2)$, $D(6, 8)$ and join *AB*, *BC*, *CD* to make a letter N.

Using $(-2, -2)$ as centre, enlarge the *N*, using a scale factor of $-\frac{1}{2}$. Label the images of *A B C D* as *A′ B′ C′ D′*.

17 Drawing axes as in question 16, plot the points $(5, -3)$ $(6, -4)$ $5, -6)$ $(4, -4)$ and join them to form a kite. This is enlarged, using a certain centre and scale factor, to form a kite with vertices at $(-4, 0)$ $(-6, 2)$ $(-4, 6)$ $(-2, 2)$. Plot this on your graph paper.

What is the scale factor of the enlargement? Join corresponding vertices of the two kites and so find the centre of enlargement.

18 On graph paper, draw the axes with values from -6 to $+6$. Plot the points $A(-2, 2)$, $B(-1, 2)$, $C(-1, -2)$, $D(-2, -2)$ and join to form a rectangle.

Using $(-4, 0)$ as centre, enlarge the rectangle using a scale factor of $+3$. Label corresponding vertices *A′ B′ C′ D′*.

The larger rectangle is also an enlargement of *ABCD* using a scale factor of -3 and a different centre of enlargement. State which vertex of the larger rectangle would be the image of *A*. What are the images of *B*, *C* and *D*? Find the centre of enlargement.

1 On graph paper draw a rectangle which measures 2 cm × 1 cm. Make a border, 1 cm wide, to surround it. This will give you a second rectangle, 4 cm × 3 cm.

Repeat this until you have four rectangles. Are they similar to one another? Are any two of them similar to one another?

If you constructed more rectangles in this manner, would any be similar to the original one?

2 Draw a rectangle which measures 6 cm × 4 cm. Draw two more rectangles similar to this, one inside and one outside the original rectangle. What are their dimensions?

3 Each of these diagrams consists of a pair of similar triangles. Copy the diagrams and mark all the pairs of equal angles you can find.

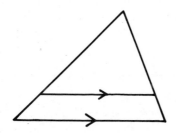
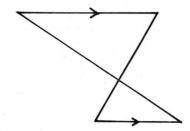

4 Both the triangles given are right angled and angle *C* = angle *Z*. Draw sketches of the two triangles and write in the given lengths.

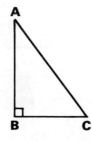
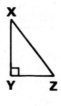

a) *BC* = 6 cm   *AB* = 8 cm   *YZ* = 1·5 cm   *ZX* = 2·5 cm
   Calculate *CA* and *XY*.

b) *AB* = 12 cm   *BC* = 5 cm   *AC* = 13 cm   *YZ* = 2·5 cm
   Calculate *XY* and *ZX*.

5 For each part of this question, draw a separate sketch of the triangles shown. Write the lengths on the diagram and do the calculations stated.

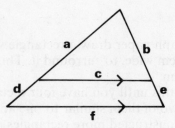

a) $a = 2$ cm $\qquad b = 2{\cdot}4$ cm $\qquad c = 2{\cdot}6$ cm $\qquad a+d = 3$ cm
Calculate $b + e$ and f.

b) $a = 3$ cm $\qquad b = 2{\cdot}4$ cm $\qquad a + d = 4$ cm $\qquad f = 3{\cdot}6$ cm
Calculate $b + e$ and c.

c) $a = 2{\cdot}5$ cm $\qquad b = 3$ cm $\qquad d = 2{\cdot}5$ cm $\qquad f = 7$ cm
Calculate e and c.

d) $a + d = 10$ cm $\qquad b + e = 15$ cm $\qquad f = 20$ cm $\qquad c = 12$ cm
Calculate a and b.

e) $a = 1{\cdot}6$ cm $\qquad b = 2{\cdot}8$ cm $\qquad c = 2$ cm $\qquad f = 5$ cm
Calculate $a + d$ and $b + e$.

6 Draw a separate sketch for each part of this question.

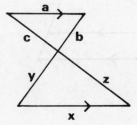

a) $a = 1{\cdot}2$ cm $\qquad b = 1{\cdot}6$ cm $\qquad c = 1{\cdot}6$ cm $\qquad y = 2{\cdot}0$ cm
Calculate x and z.

b) $x = 8$ cm $\qquad y = 8{\cdot}4$ cm $\qquad z = 7{\cdot}2$ cm $\qquad a = 2$ cm
Calculate b and c.

c) $b = 5{\cdot}7$ cm $\qquad c = 6$ cm $\qquad x = 6{\cdot}8$ cm $\qquad z = 8$ cm
Calculate a and y.

d) $x = 8{\cdot}4$ cm $\qquad y = 9{\cdot}0$ cm $\qquad z = 9{\cdot}3$ cm $\qquad a = 5{\cdot}6$ cm
Calculate b and c.

e) $a = 3$ cm $\qquad x = 10$ cm $\qquad y = 15$ cm $\qquad z = 12$ cm
Calculate b and c.

7 Draw a separate sketch for each part of this question.

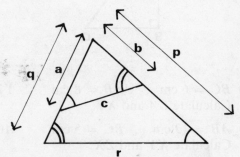

a) $a = 3.6$ cm $b = 3.9$ cm $c = 4.3$ cm $p = 7.2$ cm
   Calculate q and r.

b) $p = 12$ cm $q = 12.6$ cm $r = 10.2$ cm $b = 4.2$ cm
   Calculate a and c.

c) $a = 7.0$ cm $b = 8.4$ cm $p = 10$ cm $r = 16$ cm
   Calculate c and q.

d) $a = 6.3$ cm $c = 4.5$ cm $q = 7.2$ cm $r = 6.0$ cm
   Calculate b and p.

e) $a = 3.6$ cm $b = 4.0$ cm $q = 5.0$ cm $r = 7.5$ cm
   Calculate c and p.

8 The diagram shows an enlargement of a hexagon. Calculate the marked lengths a, b, c, d, e.

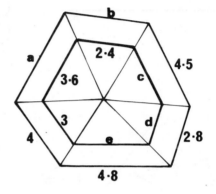

9 Two rectangles are known to be similar. The larger one measures 12 cm × 15 cm. One of the dimensions of the small one is 3 cm. What are the other possible dimensions?

10 One rectangle measures 16 cm × 48 cm. Another rectangle which is similar to the first has one measurement of 24 cm. What are its possible dimensions?

2 Reflections and Rotations

2A Reflections

1 Copy each of the diagrams and draw the images of the objects in the given mirror lines. Use tracing paper if necessary.

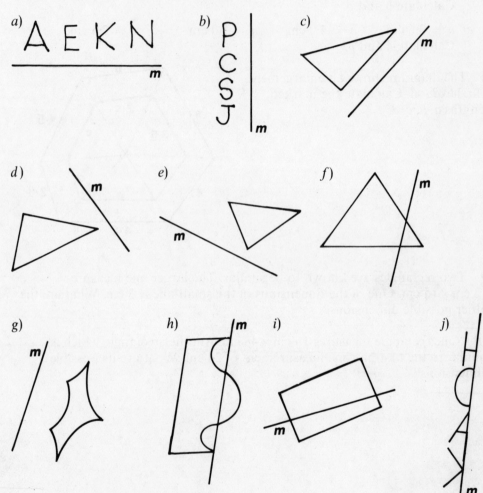

2 Write your name and then make its mirror image upside down underneath it. When you have done this, look at your 'upside down' writing in a mirror held in an upright position on the paper.

3 Plot these points: A (0·5, 0) B (0·5, 1) C (3, 1) D (2, 3) E (0·5, 4)
F (1, 6) G (0, 8).
Using $x = 0$ as the mirror lines, plot A', B', C' ..., the images of
A, B, C ....

Draw in the outline made by joining these points.
What do you notice about *a)* AB and $A'B'$, *b)* BC and $B'C'$, *c)* the angles DE and $D'E'$ make with the *y*-axis, *d)* the angles FE and $F'E'$ make with the *y*-axis?

4 Plot these points: $A (2, 1)$ $B (1, 2)$ $C (4, 3)$ $D (3, 4)$ $E (4, 5)$ $F (1, 6)$ $G (2, 7)$. Using $x = 0$ as the mirror line, plot A' B' $C' \ldots$, the images of $A, B, C \ldots$.
Join up the points. Join each point to its image.

a) What do you notice about each line joining object to image?
b) What can you say about $x = 0$ and all these lines?
c) Draw another line which is also a mirror line. What is its equation?

5 Look at the co-ordinates of the fourteen objects and images in questions 3 and 4.
Without drawing a diagram write down the images of the following points after reflection in $x = 0$.

$(1, 2)$ $(0, 5)$ $(3, 0)$ $(5, 3)$ $(2, -2)$ $(-3, 6)$ $(-5, -1)$.

6 Plot the points $A (-1.5, 0.5)$ $B (-1, 1)$ $C (0, 0)$ $D (1, 2)$ $E (2, 0)$ $F (3, 1)$ $G (3.5, 0.5)$.
Using $y = 0$ as the mirror line, find the images of these points. Join the points up and make an outline. Draw in another mirror line. What is its equation?

7 Write down the images of these points after reflection in $y = 0$.

$(2, 3)$ $(2, -3)$ $(5, 0)$ $(-4, 6)$ $(-1, -5)$ $(0, 8)$ $(3, -4)$.

8 Plot the points $A (0, 3)$ $B (1, 4)$ $C (3, 3)$ $D (-2, 2)$. Using both $y = 2$ and $x = 1$ as mirror lines, draw the images of the four points. The images in $y = 2$ are also to be reflected in $x = 1$. Plot the points so obtained. Join up all the points.
   If one of the four given points had been replaced by two other points, the figure would have had four mirror lines altogether. Which point is it?
   Replace this point by the two other points and complete the reflections in $y = 2$ and $x = 1$. Using dotted lines or colour, draw in the two new mirror lines.

9 Plot the points $A (-1, -1)$ $B (-1, 0)$ $C (1, 1)$ $D (3, 2)$ $E (3, 3)$.
Using $y = x$ as the mirror line find the images of these five points. Join up all the points.
The figure has another mirror line; draw it in. What is its equation?
Plot the points $(1, 0)$ $(2, 0)$ and $(2, 1)$ on the same axes and find their reflections in $y = x$.

10 Without drawing a diagram, write down the images of these points after reflection in $y = x$.

$(-1, 2)$ $(2, 5)$ $(3, -4)$ $(0, 4)$ $(-3, -6)$ $(7, 0)$ $(-4, 4)$

2B Rotations

In this chapter anticlockwise rotations are taken as positive.

1 Copy the flag shown into your book and, using the end *A*
as the centre of rotation, draw in the image of the flag after
it has been turned through

a) 60° in an anticlockwise direction
b) 180°
c) 90° in a clockwise direction.

2 Repeat question 1 using the points *X*, *Y* and *Z* as centres of rotation.

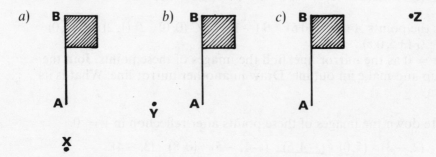

Use tracing paper to help you and draw in any other lines which you find
useful.

3 Copy triangle *ABC* into your book
and rotate it about *A* through 40° in an
anticlockwise direction.
 What angle has *AB* turned through?
What angles have *AC* and *BC* turned
through?

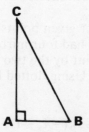

4 Copy triangle *ABC* again and this time rotate it through 40° about *B*.
What angles have the three sides turned through this time? Will this always
be true whatever the centre of rotation?

5 Copy triangle *ABC* again and this time choose a centre of rotation inside
the triangle. Turn the triangle through 50° in an anti-clockwise direction and
find what angle each side of the triangle has been rotated through.
 Will the same be true if the centre of rotation is outside the triangle?

18

6 Find the position of the centre of rotation and the angle turned through when the shaded shape is rotated on to the unshaded shape.

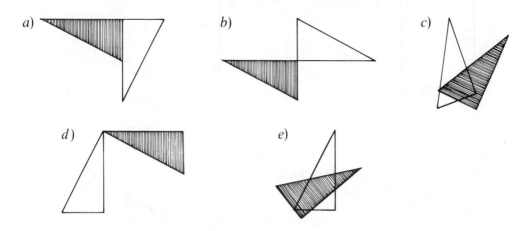

a) b) c)

d) e)

7 Find the position of the centre of rotation and the angle turned through when the shaded flag is rotated on to the unshaded one in each of these diagrams.

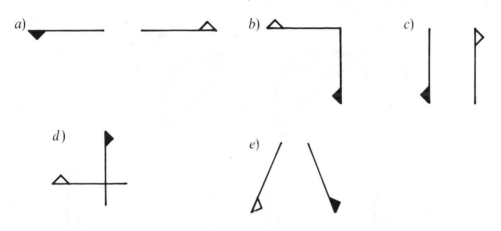

a) b) c)

d) e)

8 Copy each of the following into your book and rotate the shape about *P* in each case *i*) through 180°, *ii*) through 90°.
Mark carefully the new positions of *A, B, C, D*.

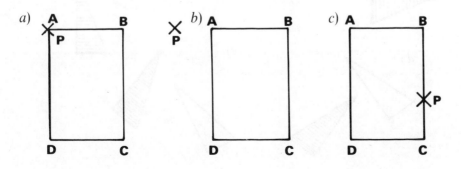

a) b) c)

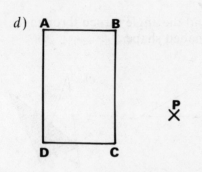

d) A   B

P
×

D   C

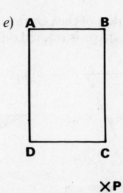

e) A   B

D   C

×P

9 Trace the diagram into your book.
Construct the mediator of *AA′* and the
mediator of *BB′* and hence find the
centre of rotation and the angle turned
through when the shaded shape is
rotated on to the unshaded.

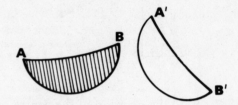

10 Repeat question 9 with each of these diagrams.

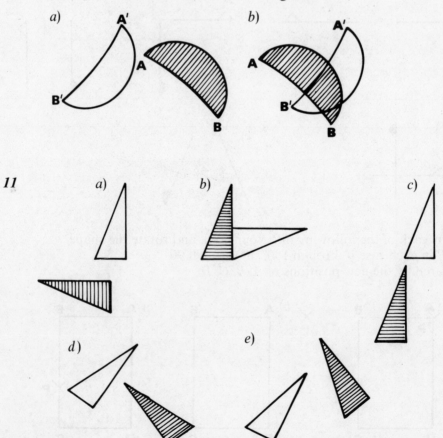

a)

b)

11   *a)*   *b)*   *c)*

d)   *e)*

Look at the diagrams on the previous page and write down those in which the shaded shape could not have been rotated on to the unshaded one.

12 On graph paper draw the triangle *T* by joining the points (1, 0) (2, 0) and (2, 2).
Mark the positions of this triangle after it has been rotated about (0, 0) through *i*) 90° *ii*) 180° *iii*) 270°. Call these triangles T_1 T_2 T_3 and the rotations R_1 R_2 R_3.

   a) How can T_1 be transformed into T_2?
   b) How can T_3 be transformed into T_2?

13   *a*) On graph paper draw the square *S* formed by joining the points
    O   *A* (2, 0)   *B* (2, 2)   and *C* (0, 2).
    Draw the square S_1 formed when *S* is rotated through 90°
    about (− 1, 0).
    Write down the co-ordinates of O_1 A_1 B_1 and C_1.
    b) S_2 with its vertices at (1, 1) (1, 3) (3, 3) and (3, 1) is the square S_1 rotated.
    Find the position of the centre of this rotation and the angle turned through.
    Label the vertices O_2 A_2 B_2 and C_2.
    c) Is there more than one possible answer to *b*)? If so, how many?
    Would the positions of the letters O_2 A_2 B_2 and C_2 change in each case?

14 It is possible to rotate the square *ABEF* on to the square *BCDE* in different ways by choosing different centres and different angles of rotation. List these and show by the order in which you letter the squares which vertex is transformed into which.

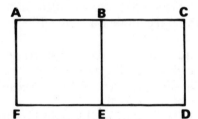

15 On graph paper draw the two triangles *ABC* and *PQR* where *A* is (1, 0) *B* (2·5, 0) *C* (2, 2) *P* (− 2·1, 1·1) *Q* (− 0·8, 0·4) and *R* (− 0·3, 2·4), leaving plenty of space above the triangles.

   By accurate construction find the centre of rotation which maps *ABC* on to *PQR*. Find also the angle turned through.

2C Miscellaneous Transformations

1 On graph paper plot the points   *A* (1, 0)   *B* (2, 0)   *C* (1, 2)   *D* (0, 4) and   *E* (0, 2). Draw the two triangles *ABC* and *ECD*.

  a) Describe one transformation that maps triangle *ABC* on to triangle *ECD*.
  b) Draw one other line to make a triangle which you can translate into triangle *ABC*. What is the translation vector?
  c) Draw a different line and make a triangle which you can reflect into triangle *ABC*. What is the mirror line?
  d) Which point would you use as the centre of enlargement to transform triangle *ECD* into triangle *OBD*?

2 On graph paper plot the points $P\,(2, 0)$ $Q\,(4, 0)$ $R\,(2, 2)$ $S\,(0, 4)$ and $T\,(0, 2)$. Join up to form the two triangles PQR and TRS.

a) Describe two transformations which map the first of these triangles on to the second. Why are there two different transformations in this case and there was only one in question 1 *a)*?

b) Draw a line and form a triangle on to which triangle PQR can be rotated. Describe the rotation.

c) How can this other triangle be transformed into triangle TRS?

3 If $ABCD$ is a parallelogram, describe a transformation which maps triangle ADC on to triangle CBA.

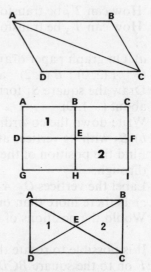

4 A rectangle is divided into four similar rectangles as shown in the diagram. Describe any transformations which will map area 1 on to area 2. Which of these will also map the whole rectangle on to itself?

5 A rectangle is divided into four triangles as shown. Describe any transformations which will map area 1 on to area 2. Which of these will also map the whole rectangle on to itself?

6 If question 4 is repeated starting with a square, there are more transformations. If question 5 is repeated starting with a square there are just the same answers. Why is this so?

7 A letter T is formed by joining the point $(2, 3)$ to $(1, 3)$, $(3, 3)$ and $(2, 1)$. Draw this T and its image T_Y formed by reflecting T in the line $y = \cdot0$. Draw also T_Z the image of T_Y after reflecting it in the line $x = 0$. Describe a single transformation which would map T on to T_Z.

8 Triangle R is formed by joining the points $(1, 1)$ $(2, 1)$ and $(1, 3)$. Draw this triangle and its image R_1 formed by reflection in the line $x = y$. Draw also R_2 the image of R_1 after reflection in the line $y = 1$. Define the transformation which maps R directly on to R_2.

9 Using the same triangle R as in question 8 and its image R_1, draw triangle R_3 formed by the reflection of R_1 in the line $x = 0$. Define the transformation which maps R directly on to R_3.

10 On graph paper plot the points $(2, 2)$ $(1, 4)$ and $(0, 3)$. Join these up and form triangle A. Translate this triangle by the vector $\binom{-2}{2}$ to position B. Rotate triangle B through $60°$ in an anticlockwise direction about the point $(-2, 1)$ on to position C.

By a ruler and compasses construction, find the centre of rotation which maps A directly on to C. What angle is turned through in this rotation?

3 Order of Operations and Simple Abstract Algebra

3A Order of Operations

$15-10$ ... 8

Find the value of each of the following expressions, working out the operation in the brackets first.

1 a) $(20+5)-9$ c) $20+(5-9)$
       b) $(20-9)+5$ d) $20-(9+5)$

4 a) $(21+6)\div3$ c) $21+(6\div3)$
       b) $(21\div3)+6$ d) $21\div(3+6)$

2 a) $(15+5)\times2$ c) $15+(5\times2)$
       b) $(15\times2)+5$ d) $15\times(2+5)$

5 a) $(3\times8)\div6$ c) $(3\div6)\times8$
       b) $3\times(8\div6)$ d) $3\div(6\times8)$

3 a) $(16-4)\times3$ c) $16-(4\times3)$
       b) $(16\times3)-4$ d) $16\times(3-4)$

6 a) $(2\times3)\times4$ c) $(2\times4)\times3$
       b) $2\times(3\times4)$ d) $2\times(4\times3)$

7 Write a list of those questions above in which two or more parts have the same answer. Note the signs in the question.

8 By convention, multiplication and division are always worked out before addition and subtraction. List those questions above in which the brackets have not made any difference, i.e. in which the same answer would have been obtained if there had been no brackets but the convention had been followed.

Complete the following by putting signs into the blanks to make them true statements:

9 $5+(6+3) = 5 \;\; 6 \;\; 3$

14 $8-2-3 = 8 \;\; (2 \;\; 3)$

10 $10-(3+2) = 10 \;\; 3 \;\; 2$

15 $9-5+2 = 9-(5 \;\; 2)$

11 $16-(10-3) = 16-10 \;\; 3$

16 $17-(10+2) = 17 \;\; 10 \;\; 2$

12 $13+5-1 = 13+(5 \;\; 1)$

17 $26-(19+3) = 26 \;\; 19 \;\; 3$

13 $12-6-1 = 12-(6 \;\; 1)$

18 $15-(9-7) = 15 \;\; 9 \;\; 7$

3B Number Sets

1 $A=\{$Positive integers from 13 to 27$\}$
Write the members of A in full.

2 $B=\{$Negative integers from -3 to $-16\}$
Write the members of B in full.

3 $C=\{x:x$ is an integer and $-5 < x < 8\}$
Write the members of C in full.

4 Remembering that zero is an integer, but neither a positive integer nor a negative integer,

if $D = \{x : x$ is a positive integer and $x \in C\}$
and $E = \{x : x$ is a negative integer and $x \in C\}$

write the members of D and E in full.

5 State the values of x, y and z if

$x = n\{$Positive integers from 11 to 40$\}$
$y = n\{$Negative integers from -6 to $-24\}$
$z = n\{$Integers between -4 and $+8\}$

6 Copy out the following statements and write against them TRUE or FALSE:

If $A = \{$Positive integers$\}$ $B = \{$Negative integers$\}$ and $C = \{$Integers$\}$

  a) $A \subset C$ *b)* $C \subset B$ *c)* $C \supset A$ *d)* $0 \in C$

  e) $-4 \notin B$ *f)* $-4 \in C$ *g)* If $D = \{-3, 2, 5\}$ $D \subset C$ and $D \not\subset A$

7 A rational number is a number of the form $\frac{p}{q}$ where p and q are integers and q is not zero. Rational numbers, in fact, are fractions, including 'top heavy' fractions such as $\frac{9}{8}$ and $\frac{6}{3}$. They are also ratios, hence the name 'rational'.

If $X = \{$All positive rationals$\}$ *a)* Give four members of X
     $Y = \{$All negative rationals$\}$ *b)* Give four members of Y
     $Z = \{$All rationals$\}$ *c)* Give four members of Z

8 *a)* What is $X \cap Y$?   *c)* What is $Y \cap Z$?
   b) What is $X \cap Z$?   *d)* What are $n(X)$, $n(Y)$ and $n(Z)$?

9 Using A and B from question 6 and X, Y and Z from question 7,

  a) Write down four numbers which are members of both A and X.
  b) Write down four numbers which are members of both B and Y.
  c) If $x \in Z$ and $x \notin Y$ and $x \notin X$, what is x?

✶   **10** If $D = \{$All rationals$\}$ $E = \{$All positive rationals$\}$ $F = \{$All negative rationals$\}$ and A, B and C have the meanings used in question 6, copy out the following statements and say whether they are true or false.

  a) $D \subset A$ *b)* $D \supset A$ *c)* $C \supset D$ *d)* $B \subset E$ *e)* $F \supset A$
  f) $F \subset A$ *g)* $E \supset A$ *h)* $E \supset B$ *i)* $E \supset F$ *j)* $E \subset F$

✶   **11** Using the same meanings for A, B, C, D, E and F as in question 10, the equation $3x + 6 = 0$ has a solution in the sets B, C, D and F but not in A or E.
State in which sets the following equations have solutions:

  a) $x - 4 = 2$ *b)* $3x + 2 = 5$ *c)* $2x + 3 = 2$
  d) $3(x + 2) = 6$ *e)* $10 - x = 7$ *f)* $x - 10 = 7$
  g) $2x - 10 = 7$

* *12* Using the same meanings for A, B, C, D, E and F as in question 10, make up equations having a solution in the sets stated. If this is not possible, say so.

   a) A, C, D and E b) B, C, D and E c) A and C only
   d) D and F only   e) C only   f) C and D only
   g) All of A, B, C, D, E and F

3C Commutative and Associative Operations

1 The order in which integers are added makes no difference to the answer. Thus $3+4 = 4+3 = 7$. Addition of integers is thus said to be commutative.
   Is the addition of rationals commutative? Give examples.

In questions *2–5*, state the answer and give examples.

2 Is the subtraction of a) integers b) rationals commutative?

3 Is the multiplication of a) integers b) rationals commutative?

4 Is division commutative? Is your answer true for all rationals and integers?

5 If A p B means 'raise A to the power of B', and if A and B are positive integers, is the operation p commutative?

6 If 's' means 'subtract the smaller of a pair of numbers from the larger', is A s B the same as B s A, i.e. is s commutative?
Give an example and explain it carefully.

7 If A z B means 'add three to A and multiply the answer by B', find the values of
   a) 4 z 3 b) 2 z 5 c) -3 z 1 d) -4 z 2
Is z commutative? Give examples.

8 If A y B means 'square A and add B' find the values of
   a) 3 y 4 b) -2 y 3 c) 4 y 1
Is y commutative? Give examples.

9 If A w B means 'raise A to the power of $2B$', find the value of
   a) 4 w 2 b) 3 w 1 c) 5 w 3
Is w commutative? Give examples.

10 The order in which three positive integers are added is immaterial, i.e. $(3+4)+5 = 3+(4+5) = 12$.
The addition of positive integers is thus said t be 'associative'.
Is the addition of all integers associative? Give examples.

In questions *11–14* state the answer and give examples.

11 Is the addition of rationals associative?

12 Is subtraction associative?
Is your answer true for all rationals and integers?

13 Is multiplication associative?
Is your answer true for all rationals and integers?

14 Is division associative?
Is your answer true for all rationals and integers?

15 There is one integer for which division fails completely when it is in the denominator. Which is it?

16 If *A z B* means 'add 2 to *A* and multiply by *B*', work out the value of

a) (3 z 2) z 5   b) (4 z 1) z 2
c) (5 z 3) z 2   d) (0 z 1) z 2

Is z associative? Give examples.

17 If *A y B* means 'square *A* and multiply by the square of *B*', work out the values of

a) (2 y 3) y 4   b) (3 y 2) y 2   c) (0 y 1) y 2

Is y associative? Give examples.

18 If *A w B* means 'take 3 from *A* and add the result to twice *B*', work out the values of

a) (4 w 1) w 2   b) (5 w 3) w 3   c) (−2 w 1) w 2

Is w associative? Give examples.

19 Are the operations z, y and w in questions 7, 8 and 9 associative?

20 Are the following pairs of operations commutative? Discuss briefly.

a) I brush my teeth and comb my hair.
b) I cut the hedge and sweep the bordering path.
c) I open my mouth and tip up my cup.
d) I address an envelope and stamp it.
e) I stamp a letter and post it.
f) I pump up a tyre and then ride my bicycle.
g) I run the water for two minutes and then put the plug in the sink.
h) I cut my leg and then bandage it.

21 Are the following arithmetic operations commutative?
Give examples and discuss.

    a) I square a number and then add 2.
    b) I multiply a number by 6 and then divide by 3.
    c) I take four from a number and then add 5.
    d) I take three from a number and then divide by 2.
    e) I square a number and then take its square root.

* 3D The Distributive Property

1   a) What is $3 \times (4+5)$?
    b) What is $(3 \times 4)+(3 \times 5)$?
    c) Are the answers to *a*) and *b*) the same?
    d) Would similar results hold for any three positive integers?

2 Your answers to 1*c*) and 1*d*) should, of course, have been YES, i.e. if
a, *b* and *c* are positive integers, $a \times (b+c)$ is equal to $(a \times b)+(a \times c)$.

    MULTIPLICATION is said to be DISTRIBUTIVE over ADDITION
    for the positive integers.

Is multiplication distributive over addition for *a*) all integers,
b) the rationals? Give examples of each.

3 Is multiplication distributive over subtraction for *a*) the integers,
b) the rationals? Give examples.

4 Is addition distributive over multiplication for the positive integers?
i.e. does $3+(4 \times 5)$ equal $(3+4) \times (3+5)$?

5 Is addition distributive over multiplication for *a*) the integers,
b) the rationals? Give examples.

6 Is multiplication distributive over division for the positive integers,
i.e. does $6 \times (12 \div 3)$ equal $(6 \times 12) \div (6 \times 3)$?

7 Is division distributive over multiplication for the positive integers?
Give examples.

8 Repeat questions 6 and 7 for the positive rationals.

9 If $A * B$ means 'square A and multiply by B', what is:

    a) $3 * 4$ b) $5 * 2$ c) $(-2) * 3$.
    d) Does $3 * (4+5)$ equal $(3 * 4) + (3 * 5)$, i.e. is $*$ distributive over
    addition for the positive integers?

10 Is addition distributive over $*$ for the positive integers? Give examples.

11 Answer questions 9 and 10 for *a*) the integers, *b*) the rationals.

12 If *A* @ *B* means 'divide *A* by the square of *B*', what is:

  a) 6 @ 4   *b*) 16 @ 2   *c*) 12 @ 3.
  d) Is the operation @ distributive over addition for the positive integers?
  Give examples.

13 Is the operation @ distributive over multiplication for the positive
integers? Give examples.

14 Is *a*) addition *b*) multiplication distributive over @ for the positive
integers? Give examples.

3E Binary Operations and Closure

1 $3+4 = 7$. The operation +, or addition, is a binary operation because
it needs two numbers at least to carry it out. Name three other binary
operations in ordinary arithmetic.

2 'Take the square root of' is not a binary operation. It concerns one number
only, e.g. $\sqrt{16} = 4$. Name three other operations in ordinary arithmetic which
are NOT binary operations.

3 Go through sections 3C and 3D and pick out all the binary operations
except the four in question 1 above. In each case give the symbol used,
the actual operation and the number of the question (since the same symbol
has been given different meanings in different questions).

E.g. 3C 6: *s* 'subtract the smaller of a pair of numbers from the larger'.

4 Are any of the operations in 3C 21 binary operations?

✱ Closure

5 3 is a positive integer. Square 3 and get 9. This is a
positive integer. When you square a positive integer, do
you always get another positive integer?

BUT, SIR, YOU SAID WE WERE GOING TO DO BINARY OPERATIONS.

6 The answer to question 5 is of course YES. The positive integers are said
to be 'closed' to the operation of squaring. Are the following sets closed to the
operation of squaring?

  a) the negative integers   *b*) the integers   *c*) the rationals
  d) the triangle numbers   *e*) the square numbers   *f*) the Fibonacci
                                                              numbers.

7 If the operation *q* means 'take the square root of', are the sets *a*) to *f*)
in question 6 closed to *q*?

8 In considering closure, it is only necessary to find one case where the set is not closed to an operation and the answer is 'not closed'. Thus the positive integers are not closed to subtraction because this sometimes gives a negative integer.

Construct a table showing whether the sets named are closed to the four basic binary operations of addition, subtraction, multiplication and division:

	Positive integers	Negative integers	Integers	Positive rationals	Negative rationals	Rationals
Addition						
Subtraction						
Multiplication						
Division			✕			✕

Why are two spaces blocked out in the last row?

9 Are the positive integers closed to the binary operations described in 3C *a*) 6 *b*) 7 *c*) 8 *d*) 16 *e*) 17 *f*) 18 and in 3D *g*) 9 *h*) 12?

10 Repeat question 9 for the negative integers.

11 Repeat question 9 for the integers.

12 In question 9*f*) and 9*h*) can you state a subset of the positive integers which is closed to the operations stated?

4 More about Numbers

4A Clock-face Arithmetic (or Residues to a Modulus)

1 Imagine a clock face marked as shown. Imagine also that it has a single hand. Set the hand on 3. Move it round a further three. Where does it finish? Set it on 2. Move it four spaces. Where does it finish? Set it on 4. Move it three spaces. Where does it finish?

2 If you got the answers right in question 1, you can now see that in this particular clock-face arithmetic $3+3 = 1, 2+4 = 1$ and $4+3 = 2$. Now find the answer to these sums using the clock face opposite:

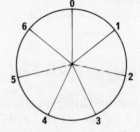

    a) $4+5$ *b*) $3+4$ *c*) $6+6$
    d) $6+6+6$ *e*) $4+4+4+4+4$

3 Sums of the type given in question 1 are examples of clock-face arithmetic to base 5. Sums of the type given in question 2 are examples of clock-face arithmetic to base 7.

    a) Five digits are used in clock-face arithmetic to base 5. What are they?
    b) Seven digits are used in clock-face arithmetic to base 7. What are they?
    c) Draw a clock face to base 6, and another to base 8. Mark on the digits.
    d) Work out the following in base 6:   *i*) $3+4$ *ii*) $1+4$ *iii*) $3+3+3$
    e) Work out the following in base 8:   *i*) $6+7$ *ii*) $5+4$ *iii*) $2+1+2$
        iv) $6+6+6$ *v*) $7+7+7+7$

4 *a*) In what way does clock-face arithmetic to base 5 differ from ordinary arithmetic to base 5?
    b) In what way does clock-face arithmetic to base 7 differ from ordinary arithmetic to base 7?
    c) A more advanced name for clock-face arithmetic is 'residues to a modulus'. In this connection the word modulus has much the same meaning as the word 'base' with which we are already familiar. In the light of your answers to *a*) and *b*) explain the meaning of 'residue'.

5 If you got the right answers to the last three of question 2, you can see that you were really multiplying. So in clock-face arithmetic to modulus 7, you now know that   *a*) $2 \times 6 = 5$ *b*) $3 \times 6 = 4$ *c*) $5 \times 4 = 6$.

Here is a complete multiplication table for clock-face arithmetic to modulus five:

	0	1	2	3	4
0	0	0	0	0	0
1	0	1	2	3	4
2	0	2	4	1	3
3	0	3	1	4	2
4	0	4	3	2	1

Construct a similar table for modulus 7.

6 Examine the two multiplication tables in question 5, one table at a time.

a) Compare the rows and columns. What do you notice?
b) Examine each row in turn to see what numbers are used in each row.
What do you notice?
c) Is a similar result true for the columns?
d) There is a line of symmetry in the table. Where is it?
Mark it on your figure. It is called the 'leading diagonal'.
e) What do you notice about the numbers in the leading diagonal?
f) Is multiplication in these two tables commutative?
(i.e. does $3 \times 2 = 2 \times 3$?)
g) Is it associative? (i.e. does $(2 \times 3) \times 4$ equal $2 \times (3 \times 4)$?)

7 *a*) Construct a multiplication table for modulus 6.
b) Which of the properties discovered in *a*) to *g*) of question 6 are still true?

✳ 8 If your table in question 7*a*) was constructed correctly, you will notice that 4×3 is 0 and also 3×2 is 0. The number zero therefore has factors, neither of which are zero.

a) In this arithmetic, could zero be the product of three factors?
Could it be the product of four factors? Give examples.
b) Are there factors of zero in the arithmetic to modulus 5? (exclude cases where 0 is a factor)
c) Are there factors of zero in arithmetic to modulus 7?

✳ 9 *a*) Without constructing the whole multiplication table, think of clock-face arithmetic to *i*) modulus 4 *ii*) modulus 8 *iii*) modulus 9
iv) modulus 11. Are there factors of zero in *i*), *ii*), *iii*) or *iv*)?
b) What do you notice about the modulus when there are NO factors of zero?
c) What do you notice about the modulus when there are factors of zero?

✳ 10 How many pairs of factors of zero are there in the arithmetic
a) to modulus 12 *b*) to modulus 24? You can answer this question without writing out the whole multiplication table. Exclude cases where zero is one of the factors, and count 3×4 and 4×3 as one set of factors.

Questions *11–19*: for these questions, the term 'clock-face arithmetic' will be discarded, and we shall talk about the arithmetic of 'residues to a modulus'.

11 *a*) Looking again at the multiplication table for residues to modulus seven, state the squares of 1, 2, 3, 4, 5, 6.
b) State the square roots of 1, 2, 3, 4, 5, 6. If there is none, say so. If there is more than one value, give all the values.

12 If you answered question 11 correctly, you now know that the square root of 1 in the arithmetic of residues to modulus 7 is either 1 or 6. This is entirely equivalent to the two square roots of 1 in ordinary arithmetic ($+1$ or -1) since 6 is equivalent to -1 in the arithmetic of residues to modulus 7.
   Does a similar result hold for the other numbers that had square roots?

13 *a*) How many numbers in the arithmetic of residues to modulus 11 have square roots?
b) How many in the arithmetic of residues to modulus 13?

   These questions can be answered without writing out the full tables. Just complete the leading diagonals.

14 Repeat question 11 for residues to modulus   *a*) 6   *b*) 8   *c*) 12.

* **15** Comparing your answer to questions 12, 13, and 14, what can you deduce about the total of numbers that have square roots in the arithmetic of
a) residues to a prime modulus   *b*) residues to a non-prime modulus?

16 Solve the equations   *a*) $x + 4 = 7$ *b*) $2x + 3 = 5$ *c*) $3x + 5 = 2$
d) $4x - 1 = 2$ *e*) $2x + 3 = 2$ in ordinary arithmetic.

17 Solve the same equations in the arithmetic of residues to modulus 11.

18 Solve the same equations in the arithmetic of residues to modulus 8.

* **19** *a*) What do you notice about ALL the answers in question 17?
b) Is the same result true for question 18?

* **20** There is only one answer for each equation in question 17 where the modulus is prime; but in question 18, where the modulus is not prime, sometimes there are two or more answers and sometimes none. By taking the modulus as 9, 10, 12, 13 and 17, see if similar results follow for other prime and non-prime moduli.

21   Division

Find the value of $3 \div 5$ (mod 7). (*Hint* Look along the 5-times line of the multiplication table till you find 3. Or write out the 5-times line till you come to 3.)
Note The suffix 'mod 7' means 'in clock-face arithmetic to modulus 7'.

22 Find the value of   *a*) $10 \div 11$ (mod 13)   *b*) $9 \div 8$ (mod 11)
c) $4 \div 3$ (mod 5)   *d*) $3 \div 4$ (mod 5)   *e*) $1 \div 6$ (mod 7)

32

23   Where possible, find the value of:

a) $6 \div 8$ (mod 12)   *b*) $4 \div 5$ (mod 6)   *c*) $5 \div 4$ (mod 6)   *d*) $3 \div 2$ (mod 5)
e) $2 \div 3$ (mod 5)   *f*) $2 \div 6$ (mod 8)   *g*) $5 \div 4$ (mod 8)   *h*) $3 \div 4$ (mod 10)

In the cases where division is impossible, what do you notice about the modulus?

✻ *24*   In the division $a \div b$ (mod *n*), why is there no answer if *a* is odd and *b* and *n* are even, but there is sometimes an answer if *a* is odd, *b* is even and *n* is odd?

4B Pascal's Triangle

1
```
              1
           1     1
        1     2     1
     1     3     3     1
  1     4     6     4     1
```

Here are the first five lines of Pascal's triangle.
Note
Any figure is obtained by adding together the two numbers above it, one slightly to the left and one slightly to the right.
Each line begins and ends with a 1. (These 1's also obey the rule given in *a*). Can you see why?)
Copy down the above five lines, and complete the table as far as the first ten lines.

2   Add up the numbers in each of the first ten rows of Pascal's triangle and record your answers as a column.

3   Looking at your answer to question 2, what would be the sum of the numbers in   *a*) row 11   *b*) row 12   *c*) row 15   *d*) row 20?

✻ *4*   Look at the 9th row beginning 1, 8, 28 etc.

The first number is 1

The second can be written as $\dfrac{8}{1}$

The third number can be written as $\dfrac{8.7}{1.2}$

The fourth number can be written as $\dfrac{8.7.6}{1.2.3}$

The fifth number can be written as $\dfrac{8.7.6.5}{1.2.3.4}$

There is no point in continuing in this way beyond the fifth number.
WHY? Using this method write down the 11th row. Check it from the tenth row.

* **5** Using the method of question 4, write down

    a) row 14 *b)* row 18 *c)* row 20.

(*Hint* 1st no.:1; 2nd no.: $\frac{14}{1} = 14$; 3rd no. = 2nd no. $\times \frac{13}{2} = 91$;

4th no. = 3rd no. $\times \frac{12}{3} = 91 \times 4 = 364$; etc.)

* **6** Check your answers to question 5 by seeing that the sum of the numbers in the rows agrees with what you found in questions 2 and 3.

7 The numbers down the outside diagonal of the triangle are all 1's. The numbers down the next diagonal column are 1, 2, 3, 4 etc. These are the natural numbers. What are the numbers in the next diagonal column (1, 3, 6, 10 etc.)?

* **8** The numbers in the 4th row (beginning 1, 3) are all divisible by 3 (except for the initial and final ones). The numbers in the 6th row are similarly divisible by 5, and in the eighth row by 7. Is this a universal rule? If not, *when* is it true?

* **9** Write the first ten rows of Pascal's triangle with the initial ones in a vertical column, thus:

    1
    1 1
    1 2 1
    1 3 3 1

By adding these numbers obliquely you obtain the Fibonacci numbers. Draw in arrows on your table to show which numbers must be added, and calculate the first few Fibonacci numbers in this way.

4C More about Fibonacci Numbers

The Fibonacci numbers are the numbers 1, 1, 2, 3, 5, 8, 13 . . . .
Each number is obtained by adding the two preceding numbers in the sequence. 5 is the fifth Fibonacci number, 8 is the sixth and 13 the seventh. So we say that 5 has index 5, 8 has index 6, 13 has index 7, etc.

1 Write down the first, third and fifth Fibonacci numbers. Add them together. You get the sixth Fibonacci number. ($1 + 2 + 5 = 8$)

    Write down the first, third, fifth and seventh Fibonacci numbers. Add them together. You get the eighth Fibonacci number. ($1 + 2 + 5 + 13 = 21$)
This is a general rule. Give a few more examples.

2 Write down the sums of the second, fourth and sixth Fibonacci numbers. Show that it is equal to the seventh number, minus 1.

    Write down the sum of the second, fourth, sixth and eighth Fibonacci numbers. Show that it is equal to the ninth Fibonacci number, minus 1. This is a general rule. Give a few more examples.

✻ *3* Call the Fibonacci numbers u_1, u_2, u_3 etc. Then u_5 is 5 and u_7 is 13. Check the truth of the following:

       8 is divisible by 4, so u_8 is divisible by u_4
       9 is divisible by 3, so u_9 is divisible by u_3
       14 is divisible by 7, so u_{14} is divisible by u_7

This is a universal rule. Write down as many examples of this as you can find.

4 The Golden Section

Divide each term of the Fibonacci sequence
by the term that precedes it.

Thus $\dfrac{1}{1} = 1$ $\dfrac{2}{1} = 2$ $\dfrac{3}{2} = 1.5$ $\dfrac{5}{3} = 1.67\ldots$

The further you go in the sequence, the nearer the value of this quotient comes to a certain limiting value. Using a calculator, slide rule or logarithms, or by dividing longhand, try and find this limiting value as accurately as you can. It is called the golden section, and buildings in which the ratio of width to height is close to this value are considered to have a pleasing appearance. Keep on dividing successive terms till the value finally settles down to a steady figure. What is this figure?

In questions 5 to 9 a number of laws about Fibonacci numbers are stated without proof. Give as many examples of them as you can.

5 A Fibonacci number is even if, and only if, its index is divisible by 3. ('Index' is defined in the note at the start of 4C.)

6 A Fibonacci number is divisible by 3 if, and only if, its index is divisible by 4.

7 A Fibonacci number is divisible by 4 if, and only if, its index is divisible by 6.

8 Try and find the conditions for divisibility by *a)* 5 *b)* 7.

9 The greatest common divisor of any two Fibonacci numbers is a Fibonacci number. Thus the greatest common divisor of 55 and 610 is 5, which is a Fibonacci number.

10 Write down the squares of the Fibonacci numbers.
Add the first four squares. They are equal to the fourth Fibonacci number multiplied by the fifth Fibonacci number:

$$1^2 + 1^2 + 2^2 + 3^2 = 1 + 1 + 4 + 9 = 15 = 3 \times 5$$

For six numbers, the sum of the squares is equal to the sixth Fibonacci number multiplied by the seventh:

$$1^2 + 1^2 + 2^2 + 3^2 + 5^2 + 8^2 = 1 + 1 + 4 + 9 + 25 + 64 = 104 = 8 \times 13.$$

Check this property for as many terms as you can manage.

11 In the Fibonacci sequence 8 lies between 5 and 13. 8^2 is 64. 5×13 is 65, i.e. one greater than 64. 13 lies between 8 and 21. 13^2 is 169. 8×21 is 168, i.e. one less than 169.

Similar results follow when you square any term of the Fibonacci sequence. Do this ten or a dozen times and try and find out when the product is one greater than the square and when it is one less.

12 Draw a fairly large regular pentagon as accurately as you can. Draw in all the diagonals.

a) Measure the sides and the diagonals. Average your measurements for the five sides and for the five diagonals. The ratio diagonal/side should be approximately equal to the golden section. (See question 4.)
b) The five diagonals enclose a second pentagon. The ratio of the side of the original pentagon to the side of the second one should be approximately equal to the golden section.
c) Draw the diagonals of the new pentagon, and repeat *a*).
d) A third pentagon now appears. Measure the side of the third pentagon as accurately as you can and compare it with the sides of the second pentagon. The ratio should be the golden section, but by now you will be losing accuracy.

The process can be continued indefinitely, but becomes very inaccurate after a time, owing to drawing errors.

4D Squares and Cubes of the Natural Numbers

1 Write the squares of the first 12 natural numbers in a column. Make a second column by writing the difference between the second and first numbers in column 1 against the second number: the difference between the third and second numbers against the third number, etc.

```
 1
 4  3
 9  5  2
16  7  2
```

Make a third column by repeating the process with each pair of numbers in the second column. What do you notice about the third column?

2 Repeat question 1 for the cubes of the first ten natural numbers. This time you will need four columns.

3 Repeat question 1 for the fourth powers of the first ten natural numbers. This time you will need five columns.

4 In a column write down the first ten natural numbers. In a parallel column write down their cumulative sums.

The columns should start:

1	1
2	3
3	6
4	10

5 In a column write down the squares of the first ten natural numbers. In a parallel column write down their cumulative sums.

The columns should start:

1	1
4	5
9	14
16	30

6 In a column write down the cubes of the first ten natural numbers. In a parallel column write down their cumulative sums.

7 In a column write down the first ten natural numbers.
In a second column, write down the numbers obtained by adding 1 to the numbers in the first column.
In the third column write down the numbers obtained by multiplying together parallel numbers in columns 1 and 2.

The columns should start:

1	2	2
2	3	6
3	4	12

Compare your last column with the second column of question 4.

* 8 In question 7, calling the numbers in the first column N and in the second column $N+1$, the numbers in the third column are $N(N+1)$. Can you deduce a formula for the sum of the first N natural numbers?

Use your formula to write down the sum of the first 15 natural numbers and also the first 20.

If a calculator is available, check your answers by actual addition.

9 In a column write down the first ten natural numbers.
In a second column write down the numbers obtained by adding 1 to the numbers in the first column.

In a third column write down the numbers obtained by doubling the numbers in the first column and adding 1.

In a fourth column write down the numbers obtained by multiplying together parallel numbers in the first three columns. Compare your answer with the second column of question 5.

* 10 In question 9, calling the numbers in the first column N, in the second column $N+1$, and in the third column $2N+1$, the numbers in the last column are $N(N+1)(2N+1)$. Can you deduce a formula for the sum of the squares of the first N natural numbers?

Use your formula to calculate the sum of the squares of the first 15 and the first 20 natural numbers. Check by actual addition if you have a calculator available.

11 Write down the first ten natural numbers in a column (column 1). In column 2 write down the numbers obtained by adding 1 to the numbers in column 1.

In column 3 write down the numbers obtained by multiplying the numbers in columns 1 and 2.

In column 4 write down the squares of the numbers in column 3.

Compare the numbers in column 4 with the numbers in column 2 of question 6. What do you notice?

* **12** Following the method of questions 8 and 10, can you deduce a formula for the sum of the cubes of the first N natural numbers?

Use your formula to find the sum of the cubes of the first 16 and the first 20 natural numbers. If you have a calculator available, check your answers by actual addition.

* **13** If you compare column 2 of questions 4 and 6, you will see that the numbers in column 2 of question 6 are the squares of the numbers in column 2 of question 4.

This means that the sum of the cubes of the first N natural numbers is the square of the sum of the numbers themselves.

Does the formula calculated in questions 8 and 12 bear this out? Check it for 6 terms, 8 terms and 10 terms.

Use this fact to write down the sums of the cubes of the first 25 natural numbers.

* **14** Write the table in question 1 in reverse:

$$
\begin{array}{ccc}
0 & 1 & 1 \\
2 & 3 & 4 \\
2 & 5 & 9
\end{array}
$$

Continue it to get the squares of the first 12 numbers.

* **15** The table in question 14 is called a difference table. Construct a difference table to give the cubes of the first 12 natural numbers. It starts:

$$
\begin{array}{cccc}
0 & 0 & 1 & 1 \\
6 & 6 & 7 & 8 \\
6 & 12 & 19 & 27 \\
6 & 18 & 37 & 64
\end{array}
$$

HA-HA! AT LAST I'VE DEDUCED A FORMULA FOR THE SUM OF THE CUBES OF THE FIRST N NATURAL NUMBERS! NOW I CAN CONQUER THE WORLD! HA-HA!!

5 Indices, Standard Form, Simple Logarithms

AAH, LOGARITHMS!

5A Indices

Work out the answers to these questions by writing each term out in full,
e.g. $a^3 = a \times a \times a$

1 $a^3 \times a^2$

2 $a^4 \times a^2 \times a$

3 $(a^3)^2$

4 $a^2b \times a^3b^2$

5 $a^2b^3c \times abc^2$

6 $a^5 \div a^2$

7 $a^4 \div a$

8 $a^4 \div a^4$

9 $a^3b^4 \div ab^3$

10 $a^2b^3 \div (ab)^2$

5B

Write down answers to the following. If you can work these out without writing each term in full, then do so.

1 $a^4 \times a^2$

2 $a^3 \times a^5$

3 $a^6 \times a$

4 $a^7 \div a^3$

5 $a^5 \div a$

6 $a^5 \times a^5$

7 $(a^5)^2$

8 $(a^2)^5$

9 $(a^4)^4$

10 $(a^8)^{\frac{1}{2}}$

11 $(a^{12})^{\frac{1}{3}}$

12 $(a^{12})^{\frac{1}{4}}$

13 $a^5 \div a^5$

14 a^0

15 $a^0 \times a \times a^2$

16 $a^2b^2 \times a^3b$

17 $(ab)^2 \times a^2$

18 $(a^4b^2)^{\frac{1}{2}}$

19 $a^3b^2 \div ab$

20 $a^4b^2 \div a^3b^2$

5C

Work out the following as simply as you can:

1 $a^4 \times a^6$

2 $a^2 \times a^7$

3 $a \times a^3 \times a^5$

4 $(a^4)^3$

5 $(a^6)^2$

6 $(a^6)^{\frac{1}{2}}$

7 $a^3b \times b^4$

8 $abc \times a^2b^3c$

9 $(ab)^3 \times a^2$

10 $(a^{12})^{\frac{1}{3}}$

11 $a^7 \div a^5$

12 $a^6 \div a^3$

13 $a^6 \div a$

14 $a^{10} \div a^{10}$

15 $ab \div b$

16 $a^4b^3 \div a^2b$

17 $a^5b^3 \div ab^2$

18 $(ab)^4 \div a^2b$

19 $a^1 \div a^2$

20 $a^5 \div a^7$

5D

Find the value of the following (leave nos. 15–20 in index form):

1 $2^2 \times 2^4$ 8 2^{-5} 15 $2^4 \times 4^3$

2 $2^3 \times 3^2$ 9 $9^{\frac{1}{2}}$ 16 $2^5 \times 4^4$

3 $3^2 \times 3^3$ 10 $9^{-\frac{1}{2}}$ 17 $2^4 \times 8^3$

4 5×5^2 11 $25^{\frac{1}{2}}$ 18 $3^3 \times 9^3$

5 6^3 12 $27^{\frac{1}{3}}$ 19 $3^4 \times 9^2$

6 5^{-1} 13 $10^4 \times 10^3$ 20 $10^8 \times 10^{-2}$

7 3^{-2} 14 $6^3 \times 5^3$

5E

Find the value of the following:

1 2×2^3 6 $3^4 \times 2^2$ 11 $3^2 \div 3^5$ 16 $8^{\frac{1}{3}}$

2 $2^3 \times 2^2$ 7 $5^2 \times 2^2$ 12 5^{-2} 17 $(8^{\frac{1}{3}})^2$

3 3×2^3 8 $5^3 \times 2^3$ 13 10^{-2} 18 $32^{\frac{2}{5}}$

4 $3^2 \times 5^2$ 9 $5^6 \times 2^6$ 14 $16^{\frac{1}{2}}$ 19 $4^{\frac{3}{2}}$

5 3^4 10 $2^3 \div 2^5$ 15 $(16^{\frac{1}{2}})^3$ 20 $125^{\frac{2}{3}}$

5F Standard Form

Write each of these numbers in full:

1 $2 \cdot 5 \times 10^3$ 8 $8 \cdot 16 \times 10^{-4}$ 15 $1 \cdot 92 \times 10^{-4}$

2 $1 \cdot 6 \times 10^5$ 9 $5 \cdot 05 \times 10^{-6}$ 16 $8 \cdot 9 \times 10^{-4}$

3 $3 \cdot 72 \times 10^4$ 10 $7 \cdot 6 \times 10^{-6}$ 17 $1 \cdot 07 \times 10^7$

4 $5 \cdot 15 \times 10^6$ 11 $3 \cdot 17 \times 10^4$ 18 $4 \cdot 63 \times 10^{-3}$

5 $8 \cdot 125 \times 10^6$ 12 $4 \cdot 95 \times 10^{-3}$ 19 $4 \cdot 092 \times 10^{-5}$

6 $9 \cdot 6 \times 10^{-2}$ 13 $2 \cdot 075 \times 10^7$ 20 $6 \cdot 2 \times 10^5$

7 $2 \cdot 75 \times 10^{-4}$ 14 $5 \cdot 006 \times 10^5$

5G

Write each of these numbers in standard form, i.e. $a \times 10^b$ where a lies between 1 and 10.

1	3200	*6*	6 390 000	*11*	0·000 76	*16*	1 072 000
2	62 000	*7*	707 000	*12*	0·000 030 8	*17*	0·0039
3	525 000	*8*	0·005 5	*13*	6 250 000	*18*	0·000 000 25
4	2090	*9*	0·000 062	*14*	12 000	*19*	0·000 417
5	17 000 000	*10*	0·000 125	*15*	0·0715	*20*	2 255 000 000

5H

Write the answer to each of the following in standard form:

1 $2·5 \times 10^3 \times 3·0 \times 10^4$

2 $1·8 \times 10^2 \times 4·1 \times 10^5$

3 $12 500 \times 30 000$

4 $2600 \times 12 000$

5 $3·6 \times 10^4 \times 4·0 \times 10^3$

6 $5·5 \times 10^3 \times 7·0 \times 10^3$

7 $8·3 \times 10^2 \times 1·5 \times 10^3$

8 $7500 \times 13 000$

9 $9100 \times 450 000$

10 $1·5 \times 10^{-2} \times 3·0 \times 10^5$

11 $4·2 \times 10^{-4} \times 2·0 \times 10^6$

12 $3·5 \times 10^{-2} \times 1·5 \times 10^{-1}$

13 $0·004 \times 5100$

14 $0·0032 \times 0·02$

15 $0·000 53 \times 0·07$

16 $2·6 \times 10^5 \div (2·0 \times 10^3)$

17 $4·5 \times 10^6 \div (3·0 \times 10^2)$

18 $0·000 64 \div 0·016$

19 $0·000 95 \div 0·000 001 9$

20 $0·0072 \div 12 000$

Logarithms

Notes

i In examples 5I to 5S either 3-figure or 4-figure tables may be used. See notes on individual exercises.

ii An alternative set of exercises 5I′ to 5N′ will be found after 5N. These approach logarithms from a different angle.

5I

Look up the logarithms of each of these numbers and write the number as a power of 10, e.g. $2·5 = 10^{0·398}$ (using 3-figure tables) or $10^{0·3979}$ (using 4-figure tables).

1	1·9	*2*	2·75	*3*	9·6	*4*	8·02	*5*	3·74
6	6·08	*7*	5·22	*8*	5·163	*9*	7·788	*10*	9·37

5J

Use your tables of logarithms to find the number equivalent to each of the following. Numbers 1 to 10 are for use with 3-figure tables, and numbers 11 to 20 for use with 4-figure tables, but all twenty can be used with either set of tables if the numbers are rounded off where necessary. (Round fives upwards.)

1	$10^{0 \cdot 544}$	*6*	$10^{0 \cdot 296}$	*11*	$10^{0 \cdot 4123}$	*16*	$10^{0 \cdot 1235}$
2	$10^{0 \cdot 654}$	*7*	$10^{0 \cdot 186}$	*12*	$10^{0 \cdot 1814}$	*17*	$10^{0 \cdot 4415}$
3	$10^{0 \cdot 703}$	*8*	$10^{0 \cdot 919}$	*13*	$10^{0 \cdot 2735}$	*18*	$10^{0 \cdot 9626}$
4	$10^{0 \cdot 771}$	*9*	$10^{0 \cdot 980}$	*14*	$10^{0 \cdot 6666}$	*19*	$10^{0 \cdot 5432}$
5	$10^{0 \cdot 600}$	*10*	$10^{0 \cdot 766}$	*15*	$10^{0 \cdot 0123}$	*20*	$10^{0 \cdot 1121}$

5K

Work out the following using logarithms.
Note Numbers 1 to 10 are for use with 3-figure tables, and numbers 11 to 20 for use with 4-figure tables, but all twenty can be used with either set of tables if the numbers are rounded off where necessary. (Round fives upwards.)

1	$3 \cdot 5 \times 2 \cdot 12$	*8*	$6 \cdot 33 \div 5 \cdot 07$	*15*	$5 \cdot 514 \times 1 \cdot 988$
2	$1 \cdot 96 \times 3 \cdot 42$	*9*	$4 \cdot 72 \div 1 \cdot 96$	*16*	$8 \cdot 364 \div 6 \cdot 289$
3	$4 \cdot 51 \times 2 \cdot 03$	*10*	$8 \cdot 83 \div 5 \cdot 29$	*17*	$4 \cdot 737 \div 1 \cdot 185$
4	$6 \cdot 6 \times 1 \cdot 32$	*11*	$4 \cdot 295 \times 2 \cdot 063$	*18*	$9 \cdot 536 \div 2 \cdot 873$
5	$2 \cdot 6 \times 2 \cdot 85$	*12*	$1 \cdot 258 \times 2 \cdot 968$	*19*	$6 \cdot 285 \div 1 \cdot 009$
6	$9 \cdot 7 \div 3 \cdot 1$	*13*	$7 \cdot 055 \times 1 \cdot 139$	*20*	$2 \cdot 852 \div 1 \cdot 133$
7	$8 \cdot 2 \div 6 \cdot 16$	*14*	$2 \cdot 518 \times 3 \cdot 262$		

5L

Write these numbers as powers of 10, using standard form and logarithms, e.g. $2500 = 2 \cdot 5 \times 10^3 = 10^{0 \cdot 398} \times 10^3 = 10^{3 \cdot 398}$ (using 3-figure tables) or $10^{3 \cdot 3979}$ (using 4-figure tables).

1	625	*4*	21 600	*7*	73 500	*10*	95 400
2	1250	*5*	30 000	*8*	40 450	*11*	20 400 000
3	7070	*6*	1288	*9*	86 000	*12*	715

13	126	*16*	3738	*19*	7000
14	4062	*17*	38 560	*20*	1200
15	9150	*18*	5954		

5M

Using the reverse process, write down the numbers equivalent to the following.

Note Numbers 1 to 15 are for use with 3-figure tables and numbers 16 to 30 for use with 4-figure tables. But all 30 can be used with either set of tables if the numbers are rounded off where necessary. (Round fives upwards.)

1	$10^{1.176}$	*11*	$10^{4.518}$	*21*	$10^{1.7777}$
2	$10^{1.845}$	*12*	$10^{4.445}$	*22*	$10^{4.2121}$
3	$10^{2.591}$	*13*	$10^{3.911}$	*23*	$10^{0.8654}$
4	$10^{2.853}$	*14*	$10^{6.602}$	*24*	$10^{3.9218}$
5	$10^{2.658}$	*15*	$10^{3.759}$	*25*	$10^{2.7879}$
6	$10^{1.932}$	*16*	$10^{1.1375}$	*26*	$10^{1.4423}$
7	$10^{1.404}$	*17*	$10^{1.7218}$	*27*	$10^{4.5026}$
8	$10^{2.189}$	*18*	$10^{2.2799}$	*28*	$10^{2.0123}$
9	$10^{2.895}$	*19*	$10^{2.3842}$	*29*	$10^{3.1666}$
10	$10^{3.283}$	*20*	$10^{3.4691}$	*30*	$10^{4.0008}$

5N

Work out the values of the following using logarithms.
Note If 3-figure tables are used, round fives upward where necessary.

1	2.5×19.7	*6*	$216 \div 58$	*11*	14.62×39.87	*16*	$156.2 \div 13.89$
2	42.3×26	*7*	$3400 \div 672$	*12*	101.8×2.636	*17*	$126.3 \div 44.38$
3	71.2×52.9	*8*	$81.6 \div 13.9$	*13*	45.71×39.28	*18*	$79.61 \div 62.35$
4	18.6^2	*9*	$62.3 \div 48.1$	*14*	16.65^3	*19*	$27.79 \div 3.525$
5	4.96^3	*10*	$459 \div 92.5$	*15*	249.8^2	*20*	$69.83 \div 48.36$

5I′ (Alternative to 5I)

Look up the logarithms of the following numbers:

1	1·9	*6*	6·08	*11*	2·37	*16*	1·33
2	2·75	*7*	5·22	*12*	4·56	*17*	7·0
3	9·6	*8*	5·163	*13*	8·91	*18*	8·149
4	8·02	*9*	7·788	*14*	7·222	*19*	3·142
5	3·74	*10*	9·37	*15*	1·4	*20*	5·568

5J′ (Alternative to 5J)

Here is a list of 20 logarithms. For each of these in turn find from your tables the corresponding number. (This is the reverse of 5I′.)

Note Numbers 1 to 10 are for use with 3-figure tables, and numbers 11 to 20 for use with 4-figure tables, but all twenty can be used with either set of tables if the numbers are rounded off where necessary. (Round fives upwards.)

1	0·544	*6*	0·296	*11*	0·4123	*16*	0·1235
2	0·654	*7*	0·186	*12*	0·1814	*17*	0·4415
3	0·703	*8*	0·919	*13*	0·2735	*18*	0·9626
4	0·771	*9*	0·980	*14*	0·6666	*19*	0·5432
5	0·600	*10*	0·766	*15*	0·0123	*20*	0·1121

5K′ (Alternative to 5K) Multiplication and Division Using Logarithms

To multiply 3·25 by 2·46 using 3-figure logarithms, set your work out as follows:

Numbers		Logs
3·25	$\longrightarrow$	0·512
2·56	$\longrightarrow$	0·403 (add)
8·22	$\longleftarrow$	0·915

(The arrows show whether you go from left to right or right to left)
The answer is 8·22.
For division, subtract instead of adding.
Use this method to answer questions 1 to 20 in 5K.

Note Numbers 1 to 10 are for use with 3-figure tables, and numbers 11 to 20 for use with 4-figure tables, but all twenty can be used with either set of tables if the numbers are rounded off where necessary. (Round fives upwards.)

5L′ (Alternative to 5L) Logarithms of Numbers Bigger Than 10

The logarithm of a number bigger than 10 is made up of two parts: the characteristic, which is a whole number, and the mantissa. The mantissa is a decimal, and each of the logarithms you have looked up so far was a mantissa without a characteristic.

To find the characteristic, count up the number of figures before the decimal point and subtract 1.

Thus 743 has three figures before the decimal point, so its characteristic is 2. Its mantissa is ·872, so the whole logarithm is 2·872. Now write down the logarithms of the 20 numbers given in 5L.

5M′ (Alternative to 5M)

Below is a list of 30 logarithms. Write down the corresponding numbers.

Note Numbers 1 to 15 are for use with 3-figure tables and numbers 16 to 30 for use with 4-figure tables. But all 30 can be used with either set of tables if the numbers are rounded off where necessary. (Round fives upwards.)

Example Logarithm 3·451 Mantissa ·451 Characteristic 3.
Equivalent number (using 3-figure tables) has the digits 283. As the characteristic is 3, there must be four figures before the decimal point. So the number is 2830.

1	1·176	*9*	2·895	*17*	1·7218	*25*	2·7879
2	1·845	*10*	3·283	*18*	2·2799	*26*	1·4423
3	2·591	*11*	4·518	*19*	2·3842	*27*	4·5026
4	2·853	*12*	4·445	*20*	3·4691	*28*	2·0123
5	2·658	*13*	3·911	*21*	1·7777	*29*	3·1666
6	1·932	*14*	6·602	*22*	4·2121	*30*	4·0008
7	1·404	*15*	3·759	*23*	0·8654		
8	2·189	*16*	1·1375	*24*	3·9218		

5N′ (Alternative to 5N)

Answer the questions in 5N, using the layout of 5K′.

Example To multiply 29·1 by 17·6 (using 3-figure tables),

Numbers	Logs	
29·1	1·464	
17·6	1·245 (add)	
512	2·709	answer is 512

THIS IS THE LAST OF THE ALTERNATIVE EXERCISES

5O

Evaluate:

1. $4 \cdot 2^2 \times 21 \cdot 6$

2. $\dfrac{21 \cdot 6 \times 2 \cdot 8}{35}$

3. $\dfrac{152 \times 8 \cdot 65}{75 \cdot 8}$

4. $9 \cdot 4^3 \times 8 \cdot 1^2$

5. $8 \cdot 6^2 + 3 \cdot 5^2$

6. $\dfrac{21\,500}{6 \cdot 2 \times 34 \cdot 7}$

7. $\sqrt{627}$

8. $\sqrt{\dfrac{56 \cdot 3}{7 \cdot 8}}$

9. $\sqrt[3]{7650}$

10. $\dfrac{59 \cdot 6 \times 18}{7 \cdot 5 \times 34 \cdot 1}$

5P

Evaluate:

1. $3 \cdot 9 \times 6 \cdot 9 \times 11 \cdot 4$

2. $\dfrac{49 \cdot 7}{2 \cdot 1 \times 5 \cdot 6}$

3. $\sqrt{51 \cdot 7}$

4. $2 \cdot 9^2 + 1 \cdot 6^3$

5. $\sqrt{\dfrac{215}{36 \cdot 4}}$

6. $5 \cdot 8 \times \sqrt{80}$

7. $\dfrac{10\,000}{52 \times 35}$

8. $\dfrac{41 \cdot 5 \times 236}{4962}$

9. $\sqrt[3]{895}$

10. $\dfrac{17 \cdot 3 \times 43}{68 \times 7 \cdot 6}$

5Q

Note When using 3-figure tables, round five's upward where necessary.

1. $2 \cdot 946 \times 3 \cdot 217$

2. $4 \cdot 314 \times 1 \cdot 962$

3. $5 \cdot 044 \times 1 \cdot 123$

4. $21 \cdot 14 \times 3 \cdot 092$

5. $14 \cdot 67 \times 8 \cdot 513$

6. $2 \cdot 981 \times 5 \cdot 819 \times 3 \cdot 666$

7. $16 \cdot 78^2$

8. $35 \cdot 17^3$

9. $4 \cdot 129^2 \times 7 \cdot 261$

10. $15 \cdot 58 \times 29 \cdot 62$

11. $8 \cdot 635 \div 4 \cdot 192$

12. $1 \cdot 946 \div 1 \cdot 051$

13. $14 \cdot 63 \div 2 \cdot 936$

14. $37 \cdot 56 \div 19 \cdot 25$

15. $13 \cdot 77 \div 9 \cdot 041$

16. $\dfrac{4 \cdot 314 \times 40 \cdot 49}{16 \cdot 85}$

17. $\dfrac{29 \cdot 83 \times 42 \cdot 78}{150 \cdot 6}$

18. $\dfrac{126 \cdot 5 \times 1 \cdot 83}{9 \cdot 581}$

19. $62 \cdot 3^2 \div 141 \cdot 8$

20. $18 \cdot 51^3 \div 6 \cdot 2^2$

* 5R

Note When using 3-figure tables, round fives upward where necessary.

1. $\sqrt{37 \cdot 61} \times 2 \cdot 83$

2. $2 \cdot 9^2 \times \sqrt{15 \cdot 2}$

3. $\sqrt{\dfrac{92 \cdot 63}{14 \cdot 88}}$

46

4 $25.6 \times \sqrt{\dfrac{48.6}{2.95}}$

5 $\sqrt[3]{621.8}$

6 $\sqrt[3]{2564}$

7 $\sqrt[3]{\dfrac{5525}{26.9}}$

8 $14.2 \times \sqrt[3]{65.92}$

9 $7.95 \times \sqrt[3]{\dfrac{162}{53}}$

10 $\sqrt[3]{56} + \sqrt{139.6}$

11 $\dfrac{2.095 \times 14.71}{13.02 \times 1.108}$

12 $3.5 \times \sqrt{\dfrac{148}{2.7}}$

13 $2.8^2 + 5.095^2$

14 $\sqrt{12.6 + 1.98^2}$

15 $\dfrac{17.92 \times \sqrt{159}}{3.56}$

16 $\dfrac{142.8 \times 625.2}{13.9}$

17 $58.6^3 \times 4.6$

18 $\dfrac{2.256 \times 95.5}{48.17}$

19 $\dfrac{4.268^2}{1.83}$

20 $\sqrt[4]{\dfrac{168.2}{73.1}}$

5S

Use logarithms to work out the answers to each of the following:

1   Find the area of a rectangle which measures 3·25 cm by 6·75 cm.

2   Find the area of a square of side 186·5 cm.

3   Find the volume of a cube which has edges of length 4·63 cm.

4   The area of a square is 2950 cm². Find the length of the sides correct to two significant figures.

5   The volume of a cube is 635 cm³. Find the length of the edges giving your answer correct to one decimal place.

6   A solid rectangular metal block which measures 9·85 cm by 7·35 cm by 2·05 cm is melted down and reformed into a solid cube. Find the length of the edges of the cube.

7   Find the volumes of two solid metal cubes, one with edges of 3 cm, the other with edges of 8·5 cm.
   If both of these cubes are melted down and reformed into one large cube, find the length of the edges of this cube.

8   A solid wooden block which stands on a square base of side 12·5 cm has a volume of 850 cm³. Find its height.

9   Two solid metal cubes, one of volume 125 cm³ and the other with a volume of 729 cm³, are melted down and formed into a larger cube. Find the lengths of the edges of all three cubes.

10   A pond which holds water to the same depth throughout has a base area of 3·75 m². Find the depth of water if the capacity is 2125 litres.

Miscellaneous Examples A

A1

1   Draw the x axis with values from -3 to 4, and the y axis with values from -5 to 4.

Draw the triangle ABC with vertices $A(1 \cdot 5, 3)$, $B(1, 1)$ and $C(2, 1)$. Taking $(0, 2)$ as centre, enlarge ABC using a scale factor of 2. Label the vertices A', B', C'.

ABC can also be mapped on to $A''B''C''$ by an enlargement, where A'' is $(-1 \cdot 5, -4 \cdot 5)$, B'' is $(-0 \cdot 5, -0 \cdot 5)$ and C'' is $(-2 \cdot 5, -0 \cdot 5)$. Draw this triangle What is the centre and scale factor of this second enlargement?

Find a transformation which would map $A'B'C'$ on to $A''B''C''$.

2   If $a*b$ means 'subtract b from twice a',

a) Find the values of
   i) $4*3$ ii) $2*5$ iii) $(3*2)*1$ iv) $(1*5)*2$ v) $1*(5*2)$
b) Are i) the integers, ii) the positive integers closed to $*$?
c) Is $*$ commutative? Give examples.
d) Is $*$ associative? Give an example.

3 $ABCD$ is a square of side 5 cm.

$X = \{P : P \text{ is a point inside the square}\}$
$Y = \{P : P \text{ is a point which is nearer } AB \text{ than } BC\}$
$Z = \{P : P \text{ is a point less than 3 cm from } BC\}$

Make an accurate drawing of the square and show the boundaries of the sets Y and Z. Shade $X \cap Y \cap Z$ and find its area.

4   In clock-face arithmetic, modulo 4, $2 + 3 = 1$.

+	0	1	2	3
0				
1				
2				1
3				

a) Copy and complete the addition table.
b) Make a similar table for multiplication.
c) Find, where possible, values of x such that:

   i) $x + 3 = 2 \pmod 4$ ii) $2 + x = 1 \pmod 4$
   iii) $2x = 3 \pmod 4$ iv) $3x + 2 = 1 \pmod 4$

d) Which numbers in this arithmetic have square roots? What are the square roots?

5   Use logarithms to find the area of a rectangular card 8·75 cm by 4·32 cm.

A square is cut from the card so that an area of 28·8 cm² remains. Find the length of the side of the square.

A2

1   a) Simplify the following, giving your answers as powers of 10:

   i) $10^3 \times 10^7$ ii) $10^8 \div 10^2$ iii) $\dfrac{10^3 \times 10^5}{10^4}$

b) Write these numbers in standard form:

    i) 72 000   *ii)* 1650   *iii)* 0·0475

c) If $a = 4·2 \times 10^3$ and $b = 3 \times 10^4$, find the values of the following, giving your answers in standard form:

    i) ab *ii)* $\dfrac{a}{b}$ *iii)* $a+b$

2 A closed cardboard box is 50 cm long, 30 cm wide and 20 cm deep. Sketch a net of the box. What is the total surface area of the outside of the box?

A similar box is 25 cm long. What is the total surface area of the outside of this box?

3 P is the point (2, 1) and Q is the point (0, 5).

    a) What are the images of P and Q after reflection in the line
        i) $y = 0$ *ii)* $x = 0$ *iii)* $y = x$?
    b) What are their images after rotations of 90° anticlockwise about
        i)   (0, 0)   *ii)* (0, 1)   *iii)* (1, 3)?
    c) P is reflected in the line $x = 1$. What further single transformation would map P on to Q?

4 *a)* If $x*y$ means 'write x in base y', find the values of $11*9, 10*8, 9*7, 8*6$. Find a pair of values of x and y which does not fit the pattern.
    b) Consider the sequence 5, 6, 7, 9, 12, 17, ...
What sequence is formed by the differences between consecutive terms? Find the next five terms of the given sequence.

5 $A = \{$Prime numbers $x: 1 < x < 16\}$
    $B = \{$Multiples of 3 less than 16$\}$
    $C = \{$Even numbers less than 16$\}$

Copy the following statements, completing them where possible by filling the boxes with suitable letters (A, B, C) or numbers from A, B and C.

    a) $\Box \in A \cap B$ *b)* $\Box \in A \cap C$ *c)* $\Box \in A \cap B \cap C$
    d) $\{2, 4, 8\} \subset \Box$ *e)* $\Box \subset B \cap C$ *f)* $\{5, 15\} \subset \Box$

A3

1 *a)* The angles of a triangle are 35°, 40° and 105°. Another triangle has an angle of 70°. Could it be similar to the first?
    b) Two quadrilaterals have angles of 60°, 90°, 90° and 120°. Are they similar?

2 *a)* Use logarithms to find the values of:
    i) $29·6 \times 35·27$ *ii)* $\dfrac{72·82}{15·33}$ *iii)* $\sqrt{4·62 \times 7·85}$
    b) Check your answer to *a i)* by seeing it lies between 25×30 and 30×40.
    c) Find the length of the side of a cube of volume 487 cm³.

3 In a golf tournament of four rounds, the par score (expected score) for a round is 70.

a) A player scores 71, 68 and 74 in the first three rounds. What must he score in the final round so that his average is par at the end of the tournament?
b) A score of 68 is described as 2 under par. Another player finishes 1 under par after the four rounds. What was his score in the final round if he had scores of 68, 71, 72 in the first 3 rounds?

4 $a*b$ means 'square a and divide by twice b'.

a) Find   *i)* $2*1$ *ii)* $4*3$ *iii)* $1*1$ *iv)* $(6*2)*3$
b) Is $*$ distributive over addition? Give an example.
c) Is $*$ commutative? Give an example.

5 If $A, B, C \ldots$ are sets within the universal set $\mathscr{E}$, and $a, b, c \ldots$ are elements of the sets, state whether the symbols in the following statements are used correctly or incorrectly:

a) $A \in B$ *b)* $a \notin C$ *c)* $n(D) = \{5\}$ *d)* $b \cap c = \{d\}$
e) $E \cup F = \{e, f, g\}$ *f)* $G \subset H'$

A4

1 In a local newspaper, advertisements cost 63p for 10 words and 6p for each additional word. When Mr Lines advertised his car for sale, it cost him £1.59. How many words did he put in the advertisement?

2 Simplify the following:

a) $y^2 \times y^4$ *c)* $y^6 \times y^3$ *e)* $y^5 \div y$ *g)* $2^4 \times 3^2$ *i)* 6^0
b) $y \times y^2 \times y^2$ *d)* $y^7 \div y^4$ *f)* $y^5 \div y^5$ *h)* 5^{-2} *j)* $3^4 \times 9^4$

(Give your answer to *j)* in index form.)

3 If M is the operation which reflects in the mirror line $x = 0$, and N is the operation which reflects in the mirror line $x = 2$, and $NM(2, 2)$ means 'reflect $(2, 2)$ in $x = 0$ and then reflect the image in $x = 2$', find the position of the image of the following points under the operations stated:

a) $M(1, 3)$ *b)* $N(1, 3)$ *c)* $NM(1, 3)$ *d)* $MN(1, 3)$
e) $M(-2, 2)$ *f)* $N(3, -2)$ *g)* $N^2(5, -1)$ *h)* $MN(2, 0)$

4 Give the next three terms in the sequence: 7, 14, 28, 56, ...
Copy and complete this table:

Term	Number	As product of primes
1st	7	7
2nd	14	7×2
3rd	28	
4th	56	
5th		
6th		
7th		

Write down the 10th term and the nth term, as products.

5 On graph paper draw the x- and y-axes taking values of -8 to $+11$ on the x-axis and -6 to $+8$ on the y-axis.
Draw the rhombus $ABCD$ with vertices $A(5, -1)$ $B(3\frac{1}{2}, 1)$ $C(2, -1)$ and $D(3\frac{1}{2}, -3)$ and the rhombus $PQRS$ with vertices $P(-1, 2)$ $Q(-4, 6)$ $R(-7, 2)$ and $S(-4, -2)$.
 Find two centres of enlargement which would map the smaller rhombus on to the larger one. Give the scale factor in each case and the vertices which correspond to A, B, C and D.

A5

1 Work out the brackets first and hence find the values of

a) $3+(4-2)-(5-1)$ d) $6-(7-3)-(5-2)$
b) $(7 \div 14) \times (16 \div 4)$ e) $3 \times (4 \times 2) \times (2 \times 3)$
c) $5+(4+3)+(6+2)$

Are the brackets superfluous in any of the above questions, i.e. would you have had the same answer if the brackets had been omitted?

2 Work on the following using logarithms.
(If you use 3-figure tables, round fives upwards where necessary.)

a) $2 \cdot 98 \times 3 \cdot 01$ b) $45 \cdot 33 \div 6 \cdot 935$ c) $\dfrac{2250 \times 6 \cdot 5}{798 \cdot 2}$

d) $3 \cdot 6^2 + 17 \cdot 2^2$ e) $\sqrt{\dfrac{75000}{908}}$

3 If T is the translation which maps the origin on to the point $(2, 3)$ and S is the translation which maps the origin on to $(-1, 2)$, find the co-ordinates of the points on to which the origin would be mapped by these translations:

a) TS b) T^2 c) S^2 d) T^2S^2 e) TST

4 *ABCD* is the square whose co-ordinates are (2, 4), (6, 2), (8, 6) and (4, 8). *PQRS* is the square whose co-ordinates are $(-2, 6)$, $(-4, 2)$, $(-8, 4)$ and $(-6, 8)$. State 3 rotations which will map *ABCD* on to *PQRS*. In each case give the centre of rotation, the angle of rotation, and the points on to which *A* and *B* are mapped.

5

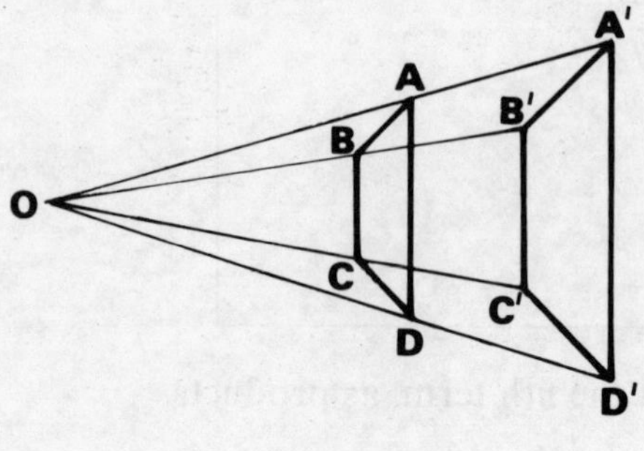

The figure shows an enlargement of *ABCD* to *A'B'C'D'* from a centre *O*. *AD* = 5 cm and *A'D'* = 7·5 cm.

a) If *OA* = 6 cm, what is *AA'*?
b) If *AA'* is 4 cm, what is *OA*?
c) If *C'D'* is 6 cm, what is *CD*?
d) If *OB'* = 10·5 cm, what is *OB*?
e) If *OC* = 8 cm, what is *OC'*?

6 Slide Rule Calculations

6A Rough Answers

A slide rule gives a numerical answer to a calculation. This answer is correct to 2 or 3 significant figures, according to the skill of the operator. It does not, however, give the position of the decimal point.

It is necessary, therefore, to be able to compute a rough answer to find where the decimal point is.

Notes on rough answers

i The rough answer should be reasonably near the true answer. If it is bigger than three times the true answer or less than one third of the true answer, this could lead to the decimal point being put in the wrong place.

ii To keep near the true answer, if you are multiplying two numbers, go up on one and down on the other. Thus, $4 \cdot 3 \times 6 \cdot 7$ could be approximated to 4×7 or to 5×6 giving either 28 or 30. The former is a better approximation.

iii If you are dividing two numbers, go up on both or down on both.

iv It is often good practice to nest a rough answer between two bounds. Thus $9 \cdot 7 \times 6 \cdot 2$ is bigger than 9×6 (54) and less than 10×7 (70). 77 squared is bigger than 70 squared (4900) and smaller than 80 squared (6400).

v If you adjust the decimal point, in multiplying two numbers move the point an equal number of places in each but in opposite directions. If dividing two numbers, move the point an equal number of places in each but in the same direction.
Thus, $473 \times 0 \cdot 052 = 4 \cdot 73 \times 5 \cdot 2$ (or approximately $5 \times 5 = 25$) but $473 \div 1240 = 4 \cdot 73 \div 12 \cdot 4$ (or approximately $4 \cdot 8 \div 12 = 0 \cdot 4$).

vi Watch out for short cuts. Thus, $0 \cdot 26 \times 51 \cdot 3$ is approximately one quarter of 52, i.e. 13.

vii A multitude of rough answers, all different, can be found for any given sum. The most useful is one that is simple to calculate but is near the true answer.

1 Do rough calculations for the following multiplications:

a) $5 \cdot 4 \times 9 \cdot 2$
b) $13 \cdot 6 \times 4 \cdot 7$
c) $49 \cdot 3 \times 87 \cdot 5$
d) $14 \cdot 2 \times 0 \cdot 376$

e) $0 \cdot 143 \times 5 \cdot 82$
f) $0 \cdot 73 \times 0 \cdot 56$
g) $1247 \times 0 \cdot 422$
h) $843 \times 0 \cdot 0661$

i) $46 \cdot 7 \times 0 \cdot 000\,29$
j) $2501 \times 0 \cdot 0682$

2 Do rough calculations for the following divisions:

a) $56.3 \div 7.91$

b) $89.4 \div 2.66$

c) $943 \div 84.2$

d) $8420 \div 17.4$

e) $\dfrac{0.0623}{7.5}$

f) $\dfrac{0.006\,14}{0.096}$

g) $\dfrac{7.21}{0.0861}$

h) $\dfrac{4324}{2.812}$

i) $\dfrac{0.0231}{79.6}$

j) $\dfrac{3142}{0.67}$

3 Do rough calculations for the following:

a) $\dfrac{4.31 \times 2.75}{5.88}$

b) $\dfrac{54.2 \times 28.6}{89.5}$

c) $\dfrac{3.18 \times 0.275}{4.43}$

d) $\dfrac{5.66 \times 0.0184}{0.0473}$

e) $\dfrac{295 \times 1044}{2833}$

4 Do rough calculations for the following:

a) $\dfrac{2.99 \times 3.61}{4.72 \times 1.85}$

b) $\dfrac{411 \times 86.2}{209 \times 41.4}$

c) $\dfrac{8.84 \times 0.275}{4.31 \times 0.552}$

d) $\dfrac{1043 \times 0.721}{28.4 \times 38.6}$

e) $\dfrac{0.773 \times 0.664}{2.21 \times 3.88}$

5 Do rough calculations for the following:

a) $\dfrac{2.73}{3.81 \times 5.26}$

b) $\dfrac{29.5}{46.2 \times 1.88}$

c) $\dfrac{0.384}{2.31 \times 8.87}$

d) $\dfrac{4.22}{96.4 \times 0.021}$

e) $\dfrac{127}{23.4 \times 0.661}$

6 Do two suitable rough calculations for each of the following sums so as to nest the answer between two limiting values:

a) 13.63×3.58

b) 46.4×0.755

c) 312×6.84

d) 4.61×5.44

e) 0.33×0.75

f) 0.437×41.4

g) 21.4×33.8

h) 741×38.6

i) 243×1216

j) 0.012×0.084

* 7 Find two limits between which the squares of the following numbers must lie:

a) 12.6 b) 8.23 c) 38.4 d) 132 e) 0.772

f) 0.361 g) 11.4 h) 148 i) 2470 j) 0.0145

* 8 Find two limits between which the square roots of the following numbers must lie:

a) 540 b) 3.9 c) 0.73 d) 2410 e) 8.14

f) $22\,800$ g) 0.47 h) 0.073 i) 149 j) 6.35

6B Slide Rule Calculations

Use your slide rule to obtain answers to the
questions in 6A. Give these answers correct to
2 significant figures. If you give a third significant
figure, put it in brackets to show it is doubtful.

1 Use the sums in question 1.

2 Use the sums in question 2.

3 Use the sums in question 3. The slide rule is particularly adapted for this
type of sum. Follow this procedure:
   Find the first number in the numerator on the D scale. Slide the number
in the denominator underneath it. Move the cursor to the second number in
the numerator on the C scale. Read what the cursor says on the D scale.
This is the answer.

4 Use the sums in question 4. The slide rule is well adapted to this type of
calculation also, and indeed to any type of calculation involving alternate
multiplication and division.

5 Use the sums in question 5.   Thus the first sum would become:
To facilitate calculation, imagine the
top has a second number, 1·0.

$$\frac{2\cdot73 \times 1\cdot0}{3\cdot81 \times 5\cdot26}$$

6 Calculate slide rule answers to the sums in question 6 and check that they
are nested by the numbers you found.

7 Find slide rule answers to the squares in question 7 and check that they
fall between the limits you found.

8 Find the square roots of the numbers in question 8. Until you are
proficient you will find it easier to pull the sliding bar right out and use the
cursor, and the A and D scales only.
(Square roots are dealt with in 14C).

In questions 9–12, first find a rough answer, and then calculate a slide rule
answer correct to two significant figures. If you calculate a third significant
figure, write it in brackets to show it is doubtful.

9

a) $\dfrac{343 \times 2\cdot76}{11\cdot9}$

b) $3\cdot46^3$

c) $0\cdot279^2$

d) $\sqrt{2320}$

e) $\dfrac{56\cdot7 \times 28\cdot2}{11\cdot9 \times 17\cdot3}$

f) $\dfrac{49\cdot1}{83 \times 21\cdot4}$

g) $22\cdot5 \times 33\cdot6 \times 44\cdot7$

h) $0\cdot123 \times 0\cdot234 \times 0\cdot345$

i) $\dfrac{37\cdot5 \times 240}{720 \times 1\cdot25}$

j) $\dfrac{1}{2\cdot34 \times 5\cdot78}$

55

* 10 Use the slide rule to express the first number in each of the following pairs as a percentage of the second. (Your answer may be bigger than 100% in some cases.)

a) 7, 11   b) 9, 16   c) 8·3, 7·5
d) 243, 417   e) 28·6, 149   f) 0·37, 1·25
g) 2·91, 0·84   h) 1240, 3870   i) 694, 921
j) 2·37, 0·84

(Percentages are dealt with in section 11)

* 11 Use the slide rule to evaluate the following:

a) 17% of 39·4   f) 32·5% of 2·88
b) 28% of 233   g) 73% of 68·2
c) 145% of 0·88   h) 0·34% of 681
d) 83% of 2145   i) 1·5% of 0·0426
e) 7% of 196   j) 233% of 0·96

* 12 Use the slide rule to turn the following fractions to decimals.

a) $\frac{5}{19}$ b) $\frac{8}{124}$ c) $\frac{79}{83}$ d) $\frac{7}{125}$

e) $\frac{56}{230}$ f) $\frac{79·3}{652}$ g) $\frac{9}{11}$ h) $\frac{1}{1256}$

i) $\frac{745}{753}$ j) $\frac{29}{68}$

7 Combining Row and Column Vectors

A vector can be represented by a) an arrow of given length and direction, or b) a column or row of 2, 3, 4, 5 . . . numbers.
a) and b) are closely connected and both have a geometrical meaning. In this chapter the word 'vector' will be used to denote a column or row of numbers only, and the geometrical meaning will be hidden or absent.

7A

1 Three schools met in a triangular athletics match. There were 10 events and points were awarded for the first three places.

School *A* had 3 first places, 5 seconds and 2 thirds.
School *B* had 1 first only, 3 seconds and 6 thirds.
School *C* had 6 firsts, 2 seconds and 2 thirds.

This information could be shown much more clearly in tabular form:

School	No. of 1sts	No. of 2nds	No. of 3rds
A	3	5	2
B	1	3	6
C	6	2	2

If the points awarded were 3 for a 1st, 2 for a 2nd and 1 for a 3rd, it is possible to work out the total scored by school *A* by setting the points down as a column vector, and preceding this by the top row of the table above (written as a row vector):

$$(3 \quad 5 \quad 2) \begin{pmatrix} 3 \\ 2 \\ 1 \end{pmatrix}$$

These two vectors 'multiplied' together give the total for school *A* as
$3 \times 3 + 5 \times 2 + 2 \times 1 = 21$.
Work out the totals for the other two schools in the same way. Which school won?

Note on vector multiplication

'Vector multiplication' is a process mathematicians have invented for combining a row and a column vector. It is convenient to call this process 'multiplication'. Here are the rules.

1   The row must precede the column.
2   The individual numbers in the row and column are combined in the manner already illustrated in question 1.
3   The row and the column must both contain the same number of 'numbers' or the process is impossible.
4   The *vector product* is a single number, not a row or column.

2 Find the product of each of the following pairs of vectors:

a) $(2 \quad 1 \quad 5) \begin{pmatrix} 3 \\ 0 \\ 1 \end{pmatrix}$

b) $(1 \quad 7) \begin{pmatrix} 5 \\ 3 \end{pmatrix}$

c) $(2 \quad -1 \quad 3) \begin{pmatrix} 2 \\ 4 \\ 1 \end{pmatrix}$

d) $(1 \quad 2 \quad -1 \quad -2) \begin{pmatrix} -3 \\ -1 \\ 2 \\ 4 \end{pmatrix}$

e) $(3 \quad -3 \quad 5) \begin{pmatrix} 2 \\ -4 \\ 1 \end{pmatrix}$

f) $(0 \quad 4 \quad 1) \begin{pmatrix} -1 \\ -3 \\ 2 \end{pmatrix}$

g) $(0 \quad 1 \quad 3 \quad -2) \begin{pmatrix} 5 \\ -2 \\ 0 \\ 6 \end{pmatrix}$

h) $(-1 \quad 2) \begin{pmatrix} 2 \\ -5 \end{pmatrix}$

i) $(3 \quad -5 \quad -1) \begin{pmatrix} -2 \\ -1 \\ 4 \end{pmatrix}$

3 Some of the vector multiplications below are impossible. If so, say why. If possible, calculate the answer.

a) $(2 \quad 3 \quad -1) \begin{pmatrix} -4 \\ 2 \end{pmatrix}$

b) $(2 \quad -4) \begin{pmatrix} 0 \\ 1 \\ 2 \end{pmatrix}$

c) $(2 \quad -1 \quad 4) \begin{pmatrix} 0 \\ -1 \\ 2 \end{pmatrix}$

d) $(0 \quad 1 \quad 2) \begin{pmatrix} -2 \\ 0 \end{pmatrix}$

e) $(-3 \quad 4 \quad 5) \begin{pmatrix} -1 \\ 2 \\ -2 \end{pmatrix}$

f) $\begin{pmatrix} 1 \\ 2 \\ 3 \end{pmatrix} (2 \quad 4 \quad 6)$

g) $\begin{pmatrix} 2 \\ 2 \end{pmatrix} (3 \quad 1 \quad 2)$

h) $(3 \quad 4) \begin{pmatrix} 3 \\ 4 \end{pmatrix}$

i) $(1 \quad 2 \quad -7) \begin{pmatrix} 3 \\ 2 \\ 1 \end{pmatrix}$

j) $(2 \quad 3 \quad -4) \begin{pmatrix} 3 \\ 4 \\ 5 \end{pmatrix}$

4 State whether the answers to the following vector multiplications are right or wrong. If wrong, give the correct answer.

a) $(3 \quad 1 \quad 2) \begin{pmatrix} 4 \\ 2 \\ 6 \end{pmatrix} = (12 \quad 2 \quad 12)$

c) $(3 \quad 0 \quad 0) \begin{pmatrix} 0 \\ 2 \\ 4 \end{pmatrix} = 0$

b) $(4 \quad 1 \quad 2) \begin{pmatrix} -2 \\ 2 \\ 3 \end{pmatrix} = 0$

d) $(5 \quad 4) \begin{pmatrix} 4 \\ 5 \end{pmatrix} = (20 \quad 20)$

5 The results of the first five matches of the season for a school's rugby team were as follows:

	Tries	Conversions	Penalties
1st match	2	1	1
2nd match	1	1	3
3rd match	3	2	0
4th match	3	1	2
5th match	4	3	2

The number of points awarded were 4 for a try, 2 for a conversion and 3 for a penalty. Write these points down as a column vector and premultiply by a row vector containing each row of the table in turn to find the school's score in each of the 5 matches.

Example 1st match $(2 \quad 1 \quad 1) \begin{pmatrix} 4 \\ 2 \\ 3 \end{pmatrix} = 13$

6 In question 5, if the results of the opposing team were as follows, find how many points were scored against the school team in each match, and how many matches they won.

	Tries	Conversions	Penalties
1st match	1	1	2
2nd match	3	3	0
3rd match	2	1	3
4th match	2	2	1
5th match	2	2	3

$(1 \ 1 \ 3) \begin{pmatrix} 4 \\ 2 \\ 3 \end{pmatrix} =$

7 If the table in question 5 had been written with the rows and columns interchanged:

	1st match	2nd	3rd	4th	5th
Tries	2	1	3	3	4
Conversions	1	1	2	1	3
Penalties	1	3	0	2	2

how could the vector multiplication have been rearranged to give the correct total scores for each match?

8 Twenty young people on an outing stopped and bought one ice each. Five of them had an ice costing 15p, seven had an ice costing 18p, and the rest had a 20p variety. Write this information in vector form and find the total cost of the ices.

7B

All the questions in this set of examples are to be worked by vector multiplication.

1 A woman buys three dozen eggs at 55p a dozen, 8 litres of milk at 18p a litre, and 3 small cartons of single cream at 20p a carton. Express the quantities as a row vector and the prices as a column vector. Pre-multiply the column vector by the row vector. This will give a single figure (a scalar) which is the total cost in pence. Work out this cost.

2 On another day, the same woman buys 10 litres of milk, 1 large carton of double cream at 60p and 2 dozen eggs. Use vector multiplication to find the total cost.

3 If the cream has gone up to 22p and the milk to $18\frac{1}{2}$p while the eggs have remained unchanged, find the new cost for the quantities in question 1.

4 A mother buys toy soldiers for her small son. She gets 20 parachutists at 14p each, 15 commandos at 16p each and 8 Welsh Guards at 22p each. What is her total expenditure?

5 A jeweller is stock-taking. He finds he has 28 gentlemen's wrist-watches, five at £28 each, four at £35 each, seven at £40 each, and the remainder at £48 each. His stock in ladies' wrist-watches is low, and he has only five at £24 and three at £32 each. Find the total value of his stock of watches.

6 The jeweller in question 5 decides to clear his stock by having a sale, so he offers all watches at 25% reduction. At the end of the sale he is left with two only of the dearest of the gentlemen's watches and one of each kind of the ladies' watches. Write down two vector multiplications, one for his sales and the other for the value of his remaining stock at the reduced price.

7 Six youth clubs in a city decide to form a league for tennis. Each club plays one game against each other club in the league. Five points are allowed for a win, three for a draw and nothing for a loss. Here is the table of the results:

	A	B	C	D	E	F
A	—	W	L	D	W	W
B	L	—	W	W	L	D
C	W	L	–	D	D	W
D	D	L	D	—	L	W
E	L	W	D	W	—	L
F	L	D	L	L	W	—

The rows show the results obtained by the various clubs. Write down a fresh table showing the number of wins, losses and draws made by each club. Using this second table, write down a vector multiplication to give the total points scored by club *A*. Repeat for clubs *B* to *F*.

Ask your teacher to show you how these six vector multiplications could have been replaced by a single matrix multiplication.

8 Repeat question 7 allowing six points for a win, four for a draw and nothing for a loss.

9 A car factory produces four models of cars, one selling at £3500, the second at £3150, the third at £2485 and the fourth at £2025. If the daily output is ten of the first, 28 of the second, 150 of the third and 410 of the fourth, what is the total value of one day's output?

10 The head of the physical education department of a large comprehensive school orders the following equipment:

	Number required	Price per item
Hockey sticks	100	£6.20
Tennis racquets	80	£8.00
Cricket bats	20	£8.50
Gloves (pairs)	10	£7.80
Pads (pairs)	10	£6.70

a) Write down a vector multiplication for the total expenditure.
b) If there is an educational discount of 10% on the first three items but not on the last two, write down a revised vector multiplication for the total expenditure.

11 The head of the mathematics department of a school orders the following text books:

For use of	No. of copies	Price per copy
VIth form	10	£1.30
VIth form	10	£2.20
VIth form	8	80p
Vth form	80	£1.15
Vth form	40	60p

Change all the prices to decimals of a pound, and write down a vector multiplication to find the total cost of the order.

12 A young man setting up a bachelor flat stocks his food cupboard from a supermarket. Here are some of his purchases:

	No. of items	Price per item
Milk	4 tins	19p
Fruit	3 tins	26p
Fruit	8 tins	30p
Meat	4 tins	45p
Bacon	2 pkts	42p
Tea	2 pkts	38p

Write down a vector multiplication and hence find his total expenditure on the above items.

13 A householder stocks up with coal for the next winter in July at summer prices. He buys five tonnes of anthracite for his central heating boiler, at £65 a tonne. He also buys three tonnes of Phurnacite for his Aga cooker, at £69 a tonne, and 1 tonne of soft coal at £50 a tonne for an occasional open fire.

Using vector multiplication, find this total expenditure.

8 Brackets and Equations

8A Brackets

Write without brackets:

1 $2(a-b+3c)$

2 $4(3a-5b)$

3 $3(1-2a+6b)$

4 $-2(5a-4b-7c)$

5 $3(x^2-x+2)$

6 $2a(a+3b-5c)$

7 $-4a(a-3)$

8 $5a(a^2-3a-5)$

9 $2ab(a^2-3ab+5b^2)$

10 $-5a^2b(3a-7b)$

8B

Write without brackets and simplify:

1 $2(a+b)-3b$

2 $3(1+a)+5(a+b)$

3 $2(4a-3b)-3(a+b)$

4 $3(1+2a-a^2)+4a^2$

5 $5a(a-1)+3$

6 $6a(a^2+2a-2)-5a^3$

7 $3(4a-3b)-2(a+b)$

8 $6a-5(a-2)$

9 $3(a+5b)-2(3a+4b)$

10 $a(1+a)-3(a-1)$

11 $10-7(a-2)$

12 $6(2-a)-4(1-3a)$

13 $3a+2(a-2b-3c)$

14 $12b-5(a+2b-3)$

15 $4a(a-1)-7$

16 $3(7-5a)-2(10-7a)$

17 $5(a-1)-2(2a+3)$

18 $3a-2(a+2b-1)+5$

19 $6(1-5a)+5(7a-1)$

20 $3(a-4b)-4(2a+3b)$

8C

Simplify the following expressions:

1 $a-2(a-b)$

2 $a(1+a)-a^2+2$

3 $a(a+2b)-ab$

4 $a(a^2-2a-3)-a^2(a-1)$

5 $ab(1-a) - a^2(1-b)$

6 $ab - 2a(b+3)$

7 $3(a-5b) + 2(6b-1)$

8 $9(1-2a) - 5(2-5a)$

9 $a(b-3) + b(a+2)$

10 $a(b+1) - b(a+4)$

11 $a^2 - a(2-a)$

12 $3a^2 - 2a(a-5)$

13 $3a(a+1) - 2(a-1)$

14 $4a(1-a+2a^2) - 3(2a-1)$

15 $ab(b+a) - a^2b$

16 $a(a+2b-5c) - b(2a-1)$

17 $a^2(a+b) - ab(a-2)$

18 $3a - a(4-b)$

19 $7(2a-3b) - 5(2a-4b)$

20 $9(1-2a) - 3(5-6a)$

8D

If $a = 2, b = -3, c = -1$ and $d = 5$, find the values of:

1 ad

2 $ad + ac$

3 $a(d+c)$

4 $d-b$

5 $d - (2a+b)$

6 $c-b$

7 $a(c-b)$

8 $a + bc$

9 $a - d(2c-b)$

10 $b(a+c) + d(a+c)$

8E

If $p = 4, q = -2, r = 3$ and $s = -1$, find the values of :

1 $p(q+r)$

2 $p + (q+r)$

3 $p - (q+r)$

4 $p - s(q+r)$

5 $r + p(q+s)$

6 $\dfrac{r+s}{q}$

7 $\dfrac{p+2q}{s}$

8 $\dfrac{p+3r}{s}$

9 $\dfrac{p^2+q^2}{5r}$

10 $\dfrac{3(p^2+q^3)}{rs}$

Write an expression for each of the following:

1 A box of chocolates contains p plain and m milk chocolates.
How many are there in the box altogether?
What is the total number of chocolates in 10 similar boxes?
How many are there in n boxes of this kind?

2 A baker's oven has five shelves. He fills one shelf with t tin loaves and
c cottage loaves. How many loaves are there altogether on this shelf?
   How many loaves will there be on the 5 shelves of the oven if they are all
filled in a similar way?
   How many loaves can be baked in a day if the baker fills his oven n times
a day?

3 Dick cuts the grass for two neighbours each week. The first gives him £a
and the second £b. How much will he make altogether in 16 weeks?

4 Bob does two jobs each week, one paying p pence and the other £q.
How much will he earn in 16 weeks a) in pence b) in pounds?

5 Kate has 30 similar parcels to tie up with ribbon. She knows that it takes
a cm to go round the parcel and b cm to tie a bow.
How much ribbon will she need altogether?
How much ribbon will she have left if she started with 18 metres?

6 Jennifer has been given £p for her birthday. She has decided to spend it on
visits to the cinema. If it costs her b pence a time for her bus fare and c pence
a time for the cinema, how many visits can she make?

7 A loose leaf file cover costs m pence and a refill n pence. If a file cover
holds 3 refills, what is the cost of one complete file?
   How much change would Tom get from a five pound note if he bought
5 such complete files?

8 Sheila is using square-based pyramids to make a decoration for
Christmas. If it takes s cm^2 of coloured paper to cover the square base and
t cm^2 to cover a triangular face, how much coloured paper is needed for one
pyramid?
   Assuming no wastage, how much coloured paper would be left from an area
of 0·5 m^2 after Sheila has made p pyramids?

9 Mrs Brown is buying fish and chips for each of the five members of her
family. If a piece of fish costs f pence and a packet of chips p pence, how much
will Mrs Brown have left from £3?

10 If it takes s grams of wool to knit a scarf and t grams to knit a pair of
gloves, how many sets, each consisting of a scarf and a pair of gloves, can
be made from 1 kg of wool?

8G

Solve the following equations:

1. $2(x+1) = 4$
2. $3x - 2(x-2) = 4$
3. $5 + 3(x-1) = 11$
4. $5(x-3) - 3 = 2$
5. $2(3x-1) = 5(x+2)$
6. $4x = 3(x+2)$
7. $2(2x-1) = 5(x-1)$
8. $7 - 2(3+x) = 0$
9. $8x = 3(3x-1)$
10. $11x - 2(4x+3) = 0$
11. $3(x-1) = 2(2x-5)$
12. $4(3-x) - 5(4-x) = 0$
13. $8 - 3(1-x) = 14$
14. $6(2-x) = 4(3+x)$
15. $10 - 3(2-3x) = 11x$
16. $1 = 3x - 2(2x+3)$
17. $5(1-x) + 2x = 8$
18. $4 + 3(1-2x) = 8x$
19. $3(2-x) - 5(1+x) = 5$
20. $2 - 5(2-3x) = 12$

8H

Solve the following equations:

1. $3(x-2) = 6$
2. $2(x+5) = 10$
3. $3(x-1) = 2(x+2)$
4. $5(x-2) - 3(x+2) = 0$
5. $4 - 3(2x+3) = 1$
6. $7x - 2(5+3x) = 3$
7. $4(1-3x) = 2$
8. $5(2-x) - 3 = 2(2x-1)$
9. $6(3-x) = 4(x+2)$
10. $10 - 3(1-2x) = 1$
11. $4(5-3x) - 3(1+2x) = 5$
12. $8(1-2x) = 5(1-2x)$
13. $11 - 7(3-x) = 2x$
14. $6(1-x) - 8(3-x) = 0$
15. $3(2x-5) - 2(x-1) = 3$
16. $4x + 7 = 5 - 2(3x+1)$
17. $8x + 2(x+3) = 5(3x-1)$
18. $5(1-2x) - 3(2-x) = 6$
19. $9 - 4(3-5x) = 7$
20. $5x - 3(3x+1) = 9$

8I Simple Examples of Rearranging Equations

In each of the following find the value of a or an expression for a in terms of b.

1. a) $6a = 12$ b) $6a = 12b$
2. a) $8a = 2$ b) $8a = 2b$
3. a) $\frac{a}{2} = 4$ b) $\frac{a}{2} = 4b$
4. a) $\frac{a}{5} = 1$ b) $\frac{a}{5} = b$
5. a) $3a = \frac{9}{10}$ b) $3a = \frac{9b}{10}$

6. a) $4a = \frac{8}{11}$ b) $4a = \frac{8b}{11}$
7. a) $6a = \frac{3}{5}$ b) $6a = \frac{3b}{5}$
8. a) $\frac{3a}{4} = 15$ b) $\frac{3a}{4} = 15b$
9. a) $\frac{5a}{9} = 20$ b) $\frac{5a}{9} = 20b$
10. a) $\frac{4a}{7} = 12$ b) $\frac{4a}{7} = 12b$

8J

Rearrange each of the following equations so that your answer starts $b = \ldots$

1. $a = 2b$
2. $a = -3b$
3. $a = \frac{1}{3}b$
4. $a = \frac{2}{3}b$
5. $a = \frac{4}{7}b$
6. $5a = 2b$
7. $8a = -3b$
8. $2a = \frac{1}{4}b$

9 $5a = \frac{b}{7}$ **14** $\frac{2}{5}b = 3a$ **19** $\frac{a}{8} = \frac{b}{6}$ **24** $\frac{3}{4}b = \frac{6}{7}a$

10 $3a = \frac{-b}{2}$ **15** $\frac{b}{2} = \frac{a}{6}$ **20** $\frac{a}{4} = \frac{b}{10}$ **25** $\frac{5}{8}b = \frac{10}{13}a$

11 $3b = \frac{a}{8}$ **16** $\frac{b}{4} = \frac{a}{5}$ **21** $\frac{a}{2} = \frac{b}{4}$

12 $2b = \frac{a}{4}$ **17** $\frac{b}{10} = \frac{a}{5}$ **22** $\frac{2}{5}b = \frac{a}{3}$

13 $\frac{3}{4}b = 6a$ **18** $\frac{b}{5} = \frac{a}{10}$ **23** $\frac{2}{7}b = \frac{1}{2}a$

8K

Rearrange each of the following equations so that your answer starts $y = \ldots$

1 $x = 5y$ **10** $\frac{x}{2} - \frac{y}{8} = 0$ **19** $5y + 3x = 11$

2 $x + 2y = 0$ **11** $x + y = 7$ **20** $3y + 6 = 12x$

3 $3x = 7y$ **12** $3x + y = 6$ **21** $2y - 5 = 8x$

4 $2x - 6y = 0$ **13** $y - x = 2$ **22** $xy = 12$

5 $x = \frac{y}{4}$ **14** $y - 3x = 1$ **23** $x = \frac{5}{y}$

6 $x = \frac{3}{4}y$ **15** $2x - y = 9$ **24** $xy - 6 = 0$

7 $\frac{x}{2} = \frac{y}{9}$ **16** $x + 5y = 1$ **25** $xy + 9 = 0$

8 $\frac{x}{3} = \frac{y}{12}$ **17** $3y - 2x = 12$

9 $\frac{x}{14} = \frac{y}{7}$ **18** $4y - 2x = 8$

8L Simple Fractional Equations

Solve the following equations:

1 $\frac{x}{2} + \frac{x}{3} = 10$
Hint Multiply each term by
6 to give $3x + 2x = 60$
Continue in the usual way.

2 $\frac{x}{5} + \frac{x}{4} = 18$
Hint Multiply each term by
20 to give $4x + 5x = 360$
Continue in the usual way.

3 $\frac{x}{2} - \frac{x}{4} = 1$ **9** $\frac{3x}{4} + \frac{x}{6} = \frac{1}{12}$ **15** $\frac{3x}{10} = \frac{2x}{5} - 1\frac{1}{2}$

4 $\frac{x}{3} - \frac{x}{4} = 2$ **10** $\frac{5x}{6} - \frac{7x}{9} = \frac{1}{3}$ **16** $\frac{1}{2}(3x + 5) = 4$

5 $\frac{x}{2} + \frac{x}{8} = \frac{5}{16}$ **11** $2 - \frac{x}{3} = x$ **17** $\frac{1}{4}(7 - 2x) = 1$

6 $\frac{x}{4} - \frac{x}{6} = \frac{1}{3}$ **12** $\frac{3}{8} - \frac{2x}{5} = \frac{x}{10}$ **18** $\frac{1}{3} = \frac{1}{5}(2x + 3)$

7 $\frac{x}{7} + \frac{x}{2} = 4\frac{1}{2}$ **13** $\frac{x}{7} - 3 = \frac{x}{4}$ **19** $\frac{1}{6}(x - 2) = \frac{x}{2}$

8 $\frac{2x}{5} - \frac{x}{7} = 3$ **14** $\frac{x}{2} + \frac{1}{4} = \frac{5x}{8}$ **20** $\frac{1}{8}(5 - 4x) = \frac{3x}{4}$

8M Simple Inequalities

Using the symbols $>$, $\geqslant$, $<$ or $\leqslant$ write the following in a shortened form:

1 The value of x is greater than 10.

2 y is a positive number.

3 a is greater than -3 and less than 7.

4 b lies between -5 and $+1$.

5 c is at least 4.

6 d is less than or equal to -5.

7 The length of the room (l) is less than twice the width (w).

8 The cost of the strawberries (s) is more than twice the cost of the cream (c).

9 The number of rabbits in the garden this year (t) is at least four times the number that were in the garden last year (l).

10 The number of choc ices sold in a day (c) is less than half the number of vanilla ices (v).

8N

State what you can about the possible values for x in each of the following.

1 $2x > 6$

2 $5x > 40$

3 $3x < 9$

4 $2x > -4$

5 $5x < -25$

6 $x+1 > 7$

7 $3x-2 \geqslant 1$

8 $2x+1 \leqslant x+5$

9 $3+x < 3x-7$

10 $7-x < 1+x$

11 $2x \leqslant 7-x$

12 $4-3x < 10-x$

13 $6+5x > x-2$

14 $4-x \geqslant 9$

15 $2(x-1) \leqslant 14$

16 $5x-1 > x$

17 $6-3(5-x) > 0$

18 $15x-4(3x+1) < 12$

19 $\frac{x}{2} < 3$

20 $\frac{x}{3}-\frac{2}{5} \geqslant 0$

21 $2x \geqslant 3$ and $4x \leqslant 10$

22 $4x < 7$ and $3x > 2x+1$

23 $3x > -7$ and $2x+1 \leqslant x+5$

24 $3x-4 \geqslant 5$ and $2x < 8$

25 $11x-6 < 8x+3$ and $2x < 10$

26 $2x+4 > 7$ and $3x-2 \leqslant x+3$

27 $3-2x \leqslant 5$ and $3+2x < 9$

28 $3(x-2) > 2(x+4)$
and $4x-3 < 2(x+15)$

29 $\frac{x}{3}+7 \geqslant \frac{x}{2}+5$ and $\frac{2(x-2)}{3} > \frac{x+1}{2}$

30 $\frac{x}{3}-\frac{3}{4} \geqslant 0$ and $2x-5 \leqslant 0$

8O Miscellaneous

1 Simplify:

a) $4(a+3b)-3(2a-b)$ b) $5x^2+3x-2(x^2-x+1)$ c) $a^2bc \times ab^2c^3$

2 If $m = -4, n = 2$, and $p = -3$, find the value of:

a) $2n+p$ b) $m(n-p)$ c) $n^2-(m+n+p)$

3 In two years' time David will be five times as old as he was six years ago. How old is he now?

4 Solve the following equations:

a) $3-2(x+1) = 0$ b) $5(2x-1)-3(3x+2) = 3$

c) $7x-5(1+3x) = 1$ d) $\frac{5x}{9} -\frac{x}{6} = 14$

5 Rearrange the following equations in the form $x = $:

a) $y = 9-x$ b) $y = \frac{3}{10}x$ c) $y = 2x+1$ -   d) $5y = \frac{2x}{3}$

6 If x is an integer and both the following inequalities are true, write all the possible values for x:

$2x > 7$ and $3x-5 \leqslant 2x+5.$

7 Solve the equations:

a) $3(x-2)+4(x-3) = 5(x-4)+4$ c) $\frac{3(x+2)}{2} = \frac{4(x+3)}{3}$

b) $\frac{x}{2}+\frac{x}{3}+\frac{x}{4}+\frac{x}{5}+\frac{x}{6} = \frac{9x}{20}+2$ d) $\frac{2}{3}(5+x) = \frac{x-5}{4}$

$*$ *8* a) The digits of the number 27 are 2 and 7. What does the 2 represent? What does the 7 represent?
b) The digits of a number are x and y. What does the x represent? What does the y represent? What is the number?
c) The digits of 27 are reversed. What is the new number?
d) The digits x and y are reversed. What does x now represent? What does y now represent? What is the new number?
e) If the sum of the numbers in b) and d) is 33, what are the numbers?

9 Maps

9A Relations

1 The Brown family consists of mother and father, two daughters, Ann and Carolyn, two sons, David and Hugh, and Mr Brown's mother.

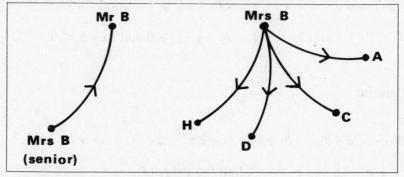

This diagram represents 'is the mother of'.
Draw similar diagrams to represent *a*) 'is the sister of'   *b*) 'is the son of'.

2 Sheila and Ken are sister and brother and are cousins of the four children in the Brown family. Draw a diagram to show the relation 'is the cousin of'.

3 Hugh and Ken go to Ashdown School, all the other children in questions 1 and 2 attend Hollypark School. Draw a diagram to show the relation 'is a pupil at', but instead of putting children and schools in one enclosure as above, have two different sets, the one with the children being the *domain* and the one with the schools being the *range*.

Ann Carolyn David Hugh Sheila Ken	Ashdown School Hollypark School

4 The six children above all have pets. Ann has a cat called Felix and a hamster called Sam. Carolyn has a dog, Rover. David also has a dog, Ben, and Hugh owns Mr Polly, a parrot. The two children in the other family have one dog, Honey, which they share, and Sheila has a tortoise, Perky. Draw a diagram to illustrate the relation 'is owned by'. List the pets in the domain and the children in the range.

5 Complete the following diagram to show the relation 'is a member of the set of':

| Square Parallelogram Isosceles triangle Rhombus Hexagon | Triangles Quadrilaterals Polygons |

6 Complete the following diagram to show the relation 'is a multiple of':

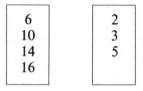

The arrows in your diagram lead from a number in the domain to one or more numbers in the range. If the arrows are reversed, what relationship is true for all of them?

Note Arrow goes from DOMAIN to RANGE. (*Mnemonic* D Depart R Arrive)

7 Look at the diagram and try to write a relationship which is true for all the pairs of numbers.

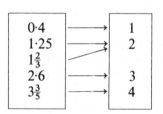

8 Write a relation which is true for the numbers linked in the diagram and complete the range.

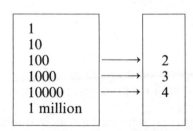

9 *a)* In the diagram, if *ABDE* is a square and *BCD* is a right-angled triangle, represent the relation 'is perpendicular to' by writing each of the six line segments *AB*, *AE*, *BD* etc. in the domain and also in the range. *b)* In a similar way, represent the relation 'is parallel to'.

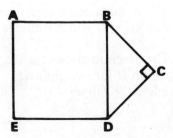

10 Complete the diagram below to illustrate the relation ' is the number of faces of':

| 4 |
| 5 |
| 6 |

| Cube |
| Tetrahedron |
| Square-based pyramid |
| Triangular prism |

9B Maps – Operations

1 If *A* is the operation which adds 1 to each member of the domain, complete the diagram below:

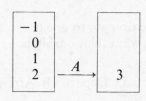

| −1 |
| 0 |
| 1 |
| 2 |

$\xrightarrow{A}$

| 3 |

2 If operation *B* is the operation which trebles each member of the domain, taking the set of numbers $\{-2, -1, 0, 1, 2\}$ to be the domain, draw the map.

3 Taking the same domain as in question 2, draw the map of operation C which squares each member of the domain.

4 Operation *D* adds 1 to the *x* co-ordinate and 2 to the *y* co-ordinate.
Complete the map for the following points:

| (0, 0) |
| (2, 0) |
| (3, 2) |
| (1, 2) |

$\xrightarrow{D}$

| (1, 2) |

5 Operation *E* changes the sign of the *x* co-ordinate. Complete the map for the same four points in the domain as in question 4.

6 For the same domain as in question 4, show the map of operation *F* which changes the sign of the *y* co-ordinate.

7 The domain consists of four points (0, 1) (0, 3) (2, 1) (2, 3). Show the map for operation *G* which doubles each number.

8 The diagram shows the map of operation *H* for two points; complete the range.

| (0, 1) |
| (0, 3) |
| (2, 1) |
| (2, 3) |

$\xrightarrow{H}$

| (−3, 0) |
| (−1, −2) |

9 Complete the range for operation I.

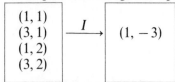

10 Operations $D, E, \ldots I$ are 'linear maps'. You should recognise them as either rotations, reflections, enlargements or translations, similar to those which you have come across already.

Plot the co-ordinates and join up the points to form closed figures in both the object and image for each map, and then describe each linear map geometrically.

Example D is a translation by the vector $\begin{pmatrix} 1 \\ 2 \end{pmatrix}$.

9C Maps – Functions

1 For the following domains complete the map:
number → prime factors of the number

a) $\{6, 10, 14, 15, 30\}$ b) $\{4, 8, 9, 16\}$

2 a) For the domain $\{1, 4, 9, 16\}$ complete the map:
number → square root of the number.

b) For the domain $\{1, 8, 27, 125\}$ complete the map:
number → cube root of the number.

3 In part b) of questions 1 and 2 the map is a function, but in part a) of each question it is not a function. Can you see the difference?

In the light of what you have noticed, state whether or not the following maps are functions:

a) Pupils in the class ⟶ their birthdays
b) Names of your friends (first name and surname) ⟶ their initials
c) Cars ⟶ registration numbers
d) Pupils in the form ⟶ their hobbies
e) Number ⟶ next highest whole number
f) Number ⟶ half the number

Now read the note that follows question 10 and reconsider your answers to question 3.

4 Using the notation $x \longrightarrow x+3$, where x is the number in the domain and $x+3$ the corresponding number in the range, complete the map:

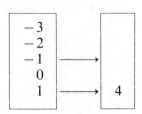

5 For the set of numbers $\{-2, -1, 0, 1, 2\}$ as the domain in each case, draw a diagram to illustrate the following maps:

a) $x \to 3x$ b) $x \to x-3$ c) $x \to 5-x$
d) $x \to 2x-3$ e) $x \to \frac{x}{4}$ f) $x \to 3x+1$

6 Taking the set of numbers $\{1, 2, 3, 4, 6\}$ as the domain in each case, draw diagrams to illustrate the following maps:

a) $x \to 2x$ b) $x \to x+2$ c) $x \to 2x+1$
d) $x \to 6-x$ e) $x \to \frac{6}{x}$

7 For each of the eleven maps in questions 5 and 6, state whether it is a function or not for the given domain.

8 For each of the same eleven maps, write down a map which would be true if all the arrows you have drawn were reversed. These are the inverses of the original maps.

9 For the map in question 6a), plot the points whose x co-ordinates are the numbers in the domain and whose y co-ordinates are the corresponding numbers in the range. Join up the points to form a straight line.

Reverse the co-ordinates, reading the range as x co-ordinates and the domain as y co-ordinates, plot these points and join up to make another line.

What do you notice about the two lines you have drawn? Repeat for 6b) and 6c). Does the same apply to these two diagrams?

10 Repeat for 6d) and 6e). In what way do these two differ from the first three?

Note A relation is a function if, and only if, each member of the domain is mapped on to one and only one member of the range.

9D Miscellaneous

1 a) If A is the relation 'has the value 0 when x is', complete the map.
b) Is A a function for the given domain?

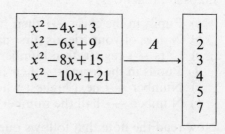

2 a) If B is the relation 'when $x = 1$, has the value', complete the map.
b) Is B a function for the given domain?

x^2-4x+3	B	0
x^2-6x+9	$\longrightarrow$	4
$x^2-8x+15$		8
$x^2-10x+21$		12

3 The first class postage rates for letters in September 1977 were:

up to 50 g	9p
over 50 g, up to 100 g	$12\frac{1}{2}$p
over 100 g, up to 150 g	16p

a) If *C* is the relation 'cost', complete the map:

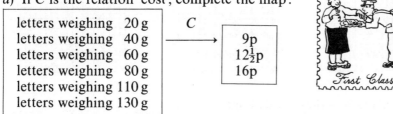

letters weighing 20 g
letters weighing 40 g
letters weighing 60 g → *C* → 9p
letters weighing 80 g $12\frac{1}{2}$p
letters weighing 110 g   16p
letters weighing 130 g

b) Is *C* a function for the given domain?
c) If the arrows were reversed, describe the new relation *C'* in words.
d) Is *C'* a function?

4 The map concerns the number of 3-nodes in various figures.
D is the relation 'has'.

a) Complete the map for the given domain. (You may rearrange the order in the domain to avoid confusion between the arrows)
b) Is *D* a function for the given domain?
c) State the inverse of *D*.
d) Is the inverse a function?

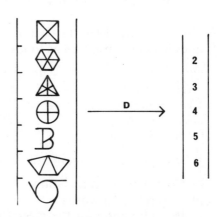

5 In a hockey match, goals were scored by the following members of the forward line: 1st goal, left wing; 2nd goal, centre forward; 3rd goal, right inner; 4th goal, right inner.

a) Draw a map to show this result, using the five members of the forward line as the domain and the four goals as the range.
b) Calling the relation *E*, describe *E* in words.
c) Is *E* a function for the given domain?
d) Is its inverse a function? Give a reason for your answer.

6 *a*) If *F* is the relation 'has a factor', complete the map.
b) Is *F* a function for the given domain and range?
c) If the arrows were reversed, calling the new relation *F'*, would *F'* be a function?

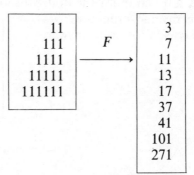

7 *a*), *b*) and *c*) are graphs of 3 functions *G*, *H* and *I*.
Copy them on to graph paper and draw the graphs of their inverses.

a) 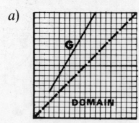   *b*) 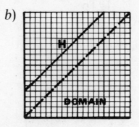   *c*)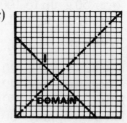

8 State the inverses of these functions:

a) $x \to x - 6$ *b*) $x \to x + 5$ *c*) $x \to 2x - 3$ *d*) $x \to 3x + 4$

e) $x \to 8 + x$ *f*) $x \to 8 - x$ *g*) $x \to \frac{x}{4}$ *h*) $x \to \frac{4}{x}$

Which of the above are their own inverses?

9 In September 1977 the cost of anthracite stove nuts delivered up to
5 miles from a depot was as follows:

Quantity	Cost per tonne
5 tonnes or over	£60
1 to 5 tonnes	£62
0·5 to 1 tonne	£64
0·1 to 0·5 tonnes	£65

Where there is ambiguity, the lower rate is charged.

a) Complete the map below.
b) Describe *J* in words.
c) Is *J* a function for the given domain?
d) Is the inverse of *J* a function?

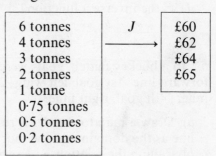

10 The graph represents a relation *K*.
a) Is *K* a function for the domain shown?
b) Is its inverse a function?
Give a reason for your answer.

76

10 Ratio and Proportion

10A Ratios

1 Write each of the following ratios in its simplest form:

16:24   25:75   49:84   2·7:6·3   1·2:14·4

2 Make the first quantity a ratio of the second:

a) 1 cm, 1 m   b) 5 cm, 1 m   c) 1 cm, 1 mm
d) 2·5 mm, 1 cm   e) 500 cm^3, 1 litre   f) 275 g, 1 kg
g) 64p, £1.20   h) £30, £2.25   i) 26p, 65p

3 Make the first quantity a ratio of the second:

a) 3·5 mm, 1·4 cm   b) 625 m, 1 km   c) 36 cm, 3 m
d) 1·4 cm, 0·35 cm   e) 500 cm^2, 1 m^2 f) 2·85 km, 3 km
g) 12 min, 1 hour   h) 2 hours, 45 min   i) 40p, £1.96

4 Express each of the following ratios in the form 1:n.

a) 1 g:1 kg   b) 1 mm^2:1 cm^2 c) 50 mm^2:1 cm^2
d) 75 mm^2:1 cm^2 e) 5p:35p   f) 5p:37·5p
g) 240 m:6 km   h) £10:£2·50   i) 210 m:140 m

5 Write each of these ratios in its simplest form:

a) £150:£450:£750   b) £2.80:£3.50:£4.20
c) 750 cm:12 m:15 m   d) 2·2 kg:0·55 kg:110 g
e) 960 cm^3:2 litres:2·4 litres

6 Find the missing quantity in each of the following:

a) 2 cm :   = 1:5   b) 2 cm: = 5:1
c)   :2 kg   = 3:8   d) 45p: = 3:11
e)   :4 litres = 9:16   f) 15 min: = 5:3
g) £600:   = 12:5   h) £7.20:   = 2:7

7 The ratio of Peter's pocket money to Paul's is 3:5. If Paul gets £1.70 a week, how much does Peter get?

8 One rectangle measures 12 cm by 8 cm and a smaller one measures 9 cm by 4 cm. What is the ratio   a) of their areas   b) of their perimeters?

9 A length of string is cut into two pieces so that one is twice as long as the other. If the shorter piece is 8 cm long, how long was the original piece of string?

10 A piece of thin wire 40 cm long is bent to form a square.
    A second piece which is twice as long is also bent to form a square. If there

is no overlapping of the ends of the pieces of wire, find the ratio of the two areas enclosed.

11 A thin wire 48 cm long is bent to form a rectangle whose width is one half of its length. What are the dimensions of the rectangle?

A second wire of the same length is bent to form a rectangle whose width is one third of its length. What are the dimensions of this rectangle?

What is the ratio of the two areas enclosed?

12 A triangle has sides whose lengths are in the ratio $4:5:6$. If the shortest side is 6 cm, how long are the other two?

13 A map is drawn on the scale of $1:50\,000$. What actual length is represented by 1 cm?

14 A plan is drawn on the scale of $1:10\,000$. What distance on the plan is equivalent to 1 km?

15 In a model of a building, an actual length of 18 m becomes 24 cm in the model. What is the scale of the model, written in the form $1:n$?

16 There are 675 pupils and 30 teachers in a certain school. Give the ratio 'number of pupils to number of teachers' in the form $n:1$.

10B Division in a Given Ratio

1 Divide £30 between two people in the ratio $1:1$, i.e. so that they have equal amounts.

2 Divide £30 between two people in the ratio $2:1$, i.e. so that the first has twice as much as the second.

3 Divide £100 between two people in the ratio $4:1$.

4 Divide a) £84 in the ratio $3:4$
b) £2.40 in the ratio $5:3$
c) £18 in the ratio $4:5$
d) £28 in the ratio $3:2$
e) £750 in the ratio $3:5$

5 Share £12 among three people in the ratio $1:2:3$, i.e. so that the largest share is equivalent to the other two put together.

6 Share £350 among three people in the ratio $2:2:1$.

7 Prize money of £14 is to be shared by two people so that one gets two and a half times as much as the other. Express this as a ratio without fractions, and find how much each receives.

8 £350 is to be shared by the first three competitors in a competition. It is to be divided so that the first prize is twice the second, and the second is twice the third. Write a suitable ratio and use it to find the three prizes.

9 A straight line 18 cm long is divided into two parts so that one part is $1\frac{1}{4}$ times the other. Find the shorter length.

10 A triangle is drawn so that the longest side is twice the shortest and the other side is $1\frac{1}{2}$ times the shortest. Find the ratio of these three lengths (without fractions), and use it to find the lengths of the three sides, if the perimeter of the triangle is 24 cm.

11 A straight line is divided into three parts 1 cm, 3 cm and 4·5 cm. Make a ratio of these three lengths using whole numbers only. What is the shorter length if the line is redivided in the ratio 2:3?

12 The perimeter of a rectangle is 33 cm. If its length and width are in the ratio 6:5, find its dimensions and area.

13 A box of chocolates contains 28 chocolates, with hard and soft centres in the ratio 3:4. After one third of the chocolates with hard centres have been eaten, and one quarter of the soft, what will be the new ratio of hard to soft?

10C Miscellaneous Ratio

1 A class is made up of 14 girls and 16 boys. *a*) Write down the ratio of girls to boys. *b*) What fraction of the school will be boys if the ratio of girls to boys in the whole school happens to be the same as in this class?
c) If the total on the school roll is 1230, how many boys are there?

2 The ratio of boys to girls in a school is 13:12. There are 650 boys. Find the number of girls.

3 Three men start a business, and their initial investments are £4800, £7200 and £8000 respectively. Over a period of 6 years their total profit is £22 500. How much should each receive if the profits are divided in the same ratio as their investments?

4 *a*) In an orchard of 72 trees, pear trees to plum trees are in the ratio of 4:5. How many plum trees are there?
   b) In the next orchard there are the same number of plum trees, but the ratio of plums to pears is 4:5. How many trees are there altogether?

5 In a factory making small machine parts, it is found that batch *A* has 400 faulty parts in a total of 20 000 and batch *B* has 270 faulty in a total of 18 000. Which is the better batch?

6 *a*) Mrs McIver makes shortbread from a recipe which requires 140 g of butter and 84 g of sugar. What is the ratio of butter to sugar?
   b) She decides next time to make a large quantity and uses 200 g of butter. How much sugar should she use?

7 In a newly opened school there are 324 girls and 243 boys. Express the ratio of boys to girls in the form $1:n$.

8 *a*) Four brothers have shares in a family business so that Arthur has 336, Bill has 224, Colin has 168 and David 112. Find the ratio of these holdings.
b) At the end of the year Arthur's share of the profits amounts to £7200. How much do the other three receive?
c) Arthur then sells his shares to his three brothers who divide them equally. Find the ratio of their new total holdings.

Nos. 9–12 should be worked by slide rule.

9 If five men can just lift a crate weighing 434 kg, how many men would be required to lift a crate weighing 563 kg? Give your answer as a whole number. Should you round up or down?

10 The area of a field is 17 000 square metres. It is divided into 3 plots whose areas are in the ratio $7:8:11$. What are the areas of each of these plots? (Answer to the nearest 100 square metres.)

11 Express the following sums of money in a ratio in the form $1:p:q$, giving your answers to two decimal places: £1143, £871, £629.

12 *a*) If Paul is 10% heavier than Peter, and John is 23% heavier than Peter, express their weights in a ratio in the form
Peter's weight : Paul's weight : John's weight $= 1:p:q$.
b) If their total weight is 221 kg, what is the weight of each (to the nearest kilogram)?

10D Proportion

Direct

1 *a*) If three books cost £1.80 and they are all sold at the same price, find the cost of one book.
b) How much would five similar books cost?
c) By using a ratio, could you have found the cost of five books without first finding the cost of one book?

2 Use a ratio to find the cost of 8 plants, knowing that 12 similar plants cost £5.40.

3 A small carton containing 125 g of plant food is sufficient for 15 plants. How many plants could be fed from a large bag holding 1 kg of the food?

4 An advertisement states that 330 ml of a new kind of varnish is sufficient to cover 22 m². How much should be needed to cover 50 m²?

5 If 56 litres of liquid cost £32, how much will 84 litres of the same liquid cost?

80

6   A car is found to travel an average distance of 100 km on 8 litres of petrol. *a)* How far should it go on 12 litres? *b)* How much petrol will be needed for a journey of 275 km?

7   A box holding 41 sweets gives the weight of the sweets and their wrappers as 205 g. How many similar sweets would you expect to find in a box giving the weight as 250 g?

8   If a man can build 8 m of wall in his garden in 12 hours, how long will it take him to put up 12 m if he works at the same rate?

Inverse

9   *a)* Mr Marsh has a stock of cigarettes which he knows will last him for 27 days if he smokes at the rate of 12 a day. How many cigarettes are in his stock?
b) If he cuts down to smoking only 8 a day, how long will his stock last?
c) You could have found this answer without finding the total number of cigarettes, by using a ratio. What ratio?

10   For the 25 cattle on his farm, Mr Thomson reckons that his stock of winter food will last for 84 days. If he increases his cattle to 35, how long will the food then last? Use a ratio to work this out.

11   A student typing at the rate of 35 words per minute took 32 minutes to copy a passage from a book. How long would it have taken a more proficient typist at a rate of 80 words per minute?

12   A school cook finds that she has sufficient potatoes to last the 360 pupils for 6 days. If the number having school dinners reduces to 270, how much longer will the potatoes last?

13   A man working 8 hours a day can paint a house in 9 days. If he only worked for 6 hours a day, how long would it take him?

14   Water from a West Country spa is exported in large plastic sterilised containers to the Middle East. Here the merchant bottles it in sterilised glass bottles each holding 450 ml. One container fills 16 of these bottles. He decides to use smaller bottles, each holding 360 ml. How many of these can he fill from one container?

15   The organiser of an outing has worked out that to cover the cost of the coach, each of the 48 passengers must pay £1.80. If 8 of them drop out, how much each must the rest pay to cover the full cost?

16   A passage from a book is copied by a typist with an average of 15 words to a line and a total length of 44 lines. Copied again with different spacing, she finds that there are only 11 words to a line. How many lines of typing will there now be?

Mixed Direct and Inverse

17 If 63 hens lay an average of 57 eggs a day, how many eggs a day would you expect from 147 hens which are kept under similar conditions?

18 (An old chestnut!) If a hen and a half lay an egg and a half in a day and a half, in how many days will 6 hens lay 12 eggs?

19 If 40 waggons, being loaded by bulldozers, can shift a pile of earth in 26 days, how long will it take to shift using 65 waggons?

20 If an aircraft flying at an average speed of 875 km/hr takes 7 hours 20 minutes to complete a journey, how long will it take for the same journey if it flies at an average speed of 1100 km/hr?

21 Using an electric fence, a farmer grazes his 74 sheep on a seeded field. He finds that they thrive when he allows them 4200 square metres a day. What total area should he allow them daily if he increases the number of sheep to 259?

22 If 46 kerb stones laid end to end stretch for 56 metres, how far will 115 of the same kerb stones stretch?

23 When there are 27 guests, a hotel chef finds he needs a delivery of potatoes every 11 days. When there are 44 guests, how often would he need a delivery of potatoes?

24 If 52 packets of 'Radiant Health Puppy Food' weigh 36 kg, what would be the weight of 117 packets?

25 A government report fills 253 pages of print, the dimensions of the rectangular block of print on each page being 7 cm by 11 cm. If these dimensions are changed to 8 cm by $11\frac{1}{2}$ cm, how many pages will now be required?

Nos. 26–30 should be calculated on a slide rule.

26 If 17 identical cakes weight 24 kg, what would be the weight of 23 cakes (to the nearest tenth of a kilogram)?

27 If a bag of meal lasts 23 hens for 17 days, how long would the same bag last 18 hens? Give your answer to the next lowest whole number of days.

28 A boy collects 43 live snails from 5 metres of his father's rockery. How many would he expect to collect from the remaining 13 metres? (Give your answer to the nearest whole snail!)

29 If his father paid the boy in question 28, 17p for clearing 5 metres of the rockery, how much should he pay him altogether for clearing the whole rockery of snails? (Answer to the nearest penny upwards.)

30 At an underground station passengers are transferred from one level to another by large lifts. If a lift can carry 38 adults whose average weight is 70 kg, without exceeding its maximum allowable weight, how many children would it carry if their average weight was 47 kg? Round your answer to the nearest whole number. Should you round upwards or downwards?

10E Miscellaneous

1 15 copies of a certain magazine cost £4.80. How much will 24 cost?

2 a) 125 small solid metal spheres each with a volume of 42 cm^3 are melted down and the metal reformed into larger solid spheres, each with a volume of 70 cm^3. How many of these spheres can be made?
b) If instead the molten metal had been remade into 350 solid spheres, all alike, what would have been the volume of each?

3 The total rail fare for 8 people on a journey is £42.40. How much will it cost for 11 people all making the same journey?

4 A supermarket has a row of cash desks down one side and a row of cold-storage units down the other. In the centre there are 44 stands of goods and these are all the same size and shape and occupy the same floor area. The total area devoted to these stands, including the necessary floor space between them, is 308 m^2. The floor area is now enlarged and an additional 56 m^2 is allocated to these stands. How many new stands can the manager install? Assuming you have done the arithmetic correctly, is your answer necessarily right? Discuss.

5 If it takes 4 men 6 days to do a job, how long will it take 3 men to do the same job, working at the same rate?

✳ **6** A carpet for a room of area 16·25 m^2 costs £104.65. How much would the same type of carpet cost for a room with an area of 14 m^2?

7 Five men set sail in a boat with enough food for 14 days. After two days, one man is lost overboard. After how many more days must the four men reach land if they do not want to go hungry?

8 Seven identical tools have a total weight of 9·1 kg. What would 12 weigh?

9 A dog breeder has enough food for his 9 dogs for a fortnight, but after one week he sells 2 of them. For how long will his supply of food now last?

10 On a day when she collects £32.40, the owner of a garden knows that she has had 72 paying visitors. How many more visitors has she had on a day when she collects £44.10?

11 If 3·8 kg of parrot food contains 110 g of grass pellets, how much would 9·5 kg contain?

12 Express the ratio 111:185:259 in its simplest form.

13 The owner of a business leaves the sum of £1500 to be divided equally among 4 employees according to their lengths of service with the firm. Mr Green had been in the firm for 23 years, Mr Brown for 24 years, Miss Poppin for 25 years and Mrs Grubb for 28 years. How much does each get?

14 If 102 packets of 'Growkwik' baby cereal weigh 36 kg, what would be the weight of 119 packets?

15 In a jubilee celebration, 28 volunteers running in relay each carry a lighted torch for a distance of 9 kilometres. How many volunteers would be required if the distance for each was reduced to 7 kilometres?

Use a slide rule in nos. 16–20

16 The total marks gained by 3 boys in an examination were 1161, 983 and 778. What is the ratio of these three sets of marks expressed in the form $1:a:b$? What is the ratio in the form $p:q:1$? (Give your answer to 2 decimal places.)

17 A kilogram of sweets is divided between five children in the ratio of their ages. Anne is 11, Peter and Susan are 12 and Elvis and Joan are 13. What weight of sweets does Anne get? (Answer to nearest 5 grams.)

18 A farmer is lifting potatoes and gives holiday work to 27 children. In one day they pick up four tonnes of potatoes. The next day he employs 32 children. How many potatoes could he expect to harvest? (Answer to the nearest tenth of a tonne.)

19 If the farmer in question 18 wishes to harvest six tonnes of potatoes a day, how many children should he employ? (Answer to the nearest whole child!)

20 Water from the bottom of a flight of locks on a canal is pumped back into the top lock to conserve the supply. If a pump delivering 4700 litres a minute can fill the lock in 55 minutes, how long would a larger pump delivering 6150 litres a minute take to fill the lock? (Answer to the nearest minute.)

Miscellaneous Examples B

B1

1 Write the following numbers to 1 s.f. and hence find rough estimates of the answers to the multiplications. Then use your slide rule to find more accurate answers.

a) $2 \cdot 2 \times 4 \cdot 1$ b) $2 \cdot 9 \times 5 \cdot 5$ c) $(6 \cdot 8)^2$
d) 23×425 e) $772 \times 4 \cdot 7$ f) $0 \cdot 45 \times 519$

2 a) If $x \rightarrow \frac{2x}{3} + 1$ find the range, given the domain $\{-3, -1, 0, 1, 3, 6\}$
     b) If $x \rightarrow 6 - x$, find the range, given the same domain as in a).

     What is the inverse function?

3 $ABCD$ is a rectangle with $AB = 7$ cm, $BC = 4$ cm.
     $X = \{$points inside the rectangle$\}$
     $Y = \{$points $P : AP < PB\}$
     $Z = \{$points $P : P$ is nearer DC than $AD\}$

Make an accurate drawing of the rectangle and mark in the boundaries of the sets Y and Z. Shade $X \cap Y \cap Z$ and find its area.

4 a) Simplify   i) $2(x+1) - (x-1)$ b) Solve the equations
           ii) $3(x+2) - 2(x-5)$ i) $2x - (x-3) = 7$
          iii) $4(2x-1) + 2(3x-3)$ ii) $4 - 3(x+1) = 5x$

5 a) Divide £1.80 between two people in the ratio $3 : 5$.
     b) A line 15 cm long is divided into three parts so that the first part is three times as long as the second, which is half as long as the third. Find the lengths of the three parts in the given order.

B2

1 a) I think of a number x, add 6 and then multiply the result by 2. Write down an expression for the final number.
     b) I start again with the same number x, but this time I multiply by 2 first and then add 6. Write down an expression for the new answer.
     c) Can you find a value of x which would give the same answer in both cases?

2 a) The supply of fresh water on a boat is sufficient to last 12 people for 20 days. If there are 15 people on board, how long will the water last? Assume that everyone uses water at the same rate.
     b) 10 men can finish a job in 32 working hours. When the job is half finished, two of the men are laid off. How long will it take the rest of them to complete the job?

3 *a)* Find *a ... e* if:

$$\begin{pmatrix} 4 \\ 3 \\ 1 \end{pmatrix} + \begin{pmatrix} 5 \\ e \\ 4 \end{pmatrix} = \begin{pmatrix} 9 \\ 3 \\ d \end{pmatrix} \text{ and } \begin{pmatrix} a \\ 0 \\ c \end{pmatrix} + \begin{pmatrix} 2 \\ b \\ 7 \end{pmatrix} = \begin{pmatrix} 5 \\ 1 \\ 12 \end{pmatrix}$$

b) Find *a, b, c* if:

$$(3 \ 0 \ a) \begin{pmatrix} 4 \\ -b \\ 2 \end{pmatrix} = 14; \quad (2 \ 1 \ 5) \begin{pmatrix} 4 \\ b \\ 2 \end{pmatrix} = 18 \text{ and } (2 \ 4 \ 1) \begin{pmatrix} 4 \\ -b \\ 2 \end{pmatrix} = c$$

4 *a)* In a rectangle, the length exceeds the width by 4 cm. If the width is x cm, write down an expression for the perimeter and draw the map width $\rightarrow$ perimeter for the domain $\{2, 4, 6, 8, 10\}$ cm.
b) What is the perimeter of a rectangle of width 3 cm?
c) Is the map a function for the given domain?
d) Is there an inverse function?
e) If your answer to *d)* was YES, find the inverse function and use it to find the dimensions of a rectangle of perimeter 30 cm.

5 Use your slide rule to find the values of:

a) $2{\cdot}4 \times 3{\cdot}7$ *b)* $16{\cdot}5 \times 9{\cdot}2$ *c)* $\frac{37 \cdot 4}{8 \cdot 29}$ *d)* $(0{\cdot}46)^2$

Give a rough check for *d*).

B3

1 In a campaign against smoking it is claimed that smoking one cigarette reduces the expectation of life by $5\frac{1}{2}$ minutes.

a) A man smokes 20 cigarettes a day for a week. If this claim is true, how much has his expectation of life been reduced?
b) He decides to come down to 10 cigarettes a day, and he keeps to this for 4 weeks. By how much, to the nearest day, has his expectation of life been further reduced?
c) If packets of 20 cigarettes cost 52p, how much would he have saved if he had stopped smoking altogether at the beginning of this 5 week period?

2 *a)* Which of the following would be 6·1 when written correct to 2 s.f.?
     5·98   6·04   6·082   6·145   6·18   6·195
b) Make rough estimates of the answers to these calculations by first writing each number correct to 1 s.f.

     i) $4{\cdot}6 \times 2{\cdot}7$ *ii)* $9{\cdot}47 \times 1{\cdot}23$ *iii)* $0{\cdot}423 \times 4{\cdot}67$
     iv) $0{\cdot}27 \div 6{\cdot}41$ *v)* $0{\cdot}027 \div 0{\cdot}323$ *vi)* $(0{\cdot}067)^2$

3 I think of a number, x, multiply it by 3 and then subtract 2.

   a) Write this as a map $x \rightarrow \ldots$, where the domain is the positive integers.
   b) Is this a function?
   c) If I thought of 5, what would be the final result?
   d) What is the inverse? Is it a function?
   e) If my final answer is 19, what number did I first think of?

4 A triangle ABC has its vertices at the points $(2, 4)$ $(2, 6)$ and $(4, 5)$. It is translated by the vector $\left(_{-\frac{3}{2}}\right)$ and then rotated through $90°$ about the origin.
   Find the position of its image and show that it could have been mapped on to the same image by a single rotation of $90°$ about P.
   Find the co-ordinates of P by a 'ruler and compass' construction.
Show your construction lines.

5 *a*) The ages of two sisters differ by 5 years. If the younger is x years old, what is the age of her sister?
   b) In 3 years' time, the older will be twice the age of her sister. Give expressions for their ages at that time. What are their actual ages now?

B4

1 Evaluate the following vector products. Where this is not possible, say so.

   a) $\begin{pmatrix} 1 \\ 2 \\ 3 \end{pmatrix}$ $(2 \quad 0 \quad 1)$ *b*) $(2 \quad -1 \quad 0) \begin{pmatrix} 2 \\ -1 \\ 0 \end{pmatrix}$ *c*) $(3 \quad 5 \quad 1) \begin{pmatrix} 2 \\ 4 \end{pmatrix}$

   d) $(0 \quad 1 \quad 0 \quad 2) \begin{pmatrix} 4 \\ -2 \\ 1 \\ -3 \end{pmatrix}$ *e*) $(-5 \quad 4) \begin{pmatrix} -2 \\ -3 \end{pmatrix}$

2 Solve the following:

   a) $6 - x = 2x$ *b*) $4 + x = 1 - 2x$ *c*) $2(x + 1) = 3(x - 2)$

   d) $6(1 + 2x) - 3(3x - 2) = 10$ *e*) $5 - 3(4 - 2x) = 4x$

3 Square 1 can be mapped on to square 2 in different ways by single transformations.
List these ways and draw a diagram for each case to show the positions of $A'B'C'D'$ (the images of A, B, C and D).

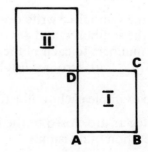

4 The diagram shows a cuboid in which the length, breadth and height are all different. Draw the map 'is equal in length to' for the domain {*AB, BC, CD, AE*}.

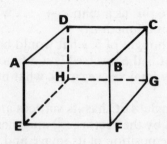

In such a cuboid, if the length is three times the breadth and the height is one quarter of the length, write a ratio for length : breadth : height. If the total length of the edges of the cuboid is 152 cm, find its dimensions.

5 For domestic use, the tariff for gas is as follows:
The first 100 therms cost 18·10 pence per therm.
The remaining therms cost 16·60 pence per therm.
There is also a standing charge of £1.00 per quarter.

a) In one quarter I used 125·42 therms and in the next quarter I used 50·71 therms. Find my bill for each of these quarters.
b) Draw a graph to show the cost of gas, plotting therms along the *x*-axis and 'cost in £' along the *y*-axis. There will be a change of slope at *x* = 100.
c) From your graph, check your answers to *a*).

B5

1 If *A* = {Integers}, *B* = {Positive integers}, *C* = {Negative integers}, *D* = {Rationals}, *E* = {Positive rationals}, *F* = {Negative rationals}, *G* = {Irrationals}, *H* = {Reals}.
(*Note H* = {*A, D, G*})

a) Using the symbol ⊂ write down 8 relations between the sets listed.
b) Using the symbols ∈ and ∉ write down the relationship of each of the following numbers to each of the sets *A* to *H*:
   i) −0·75   *ii)* zero   *iii)* 17

✱ *2* Give one example each to illustrate the following:

a) The integers are closed to the binary operation of multiplication but not to the operation of division.

b) Addition of rationals is associative.

c) For the integers, multiplication is distributive over addition, but addition is not distributive over multiplication.

d) Excluding division by zero, the rationals are closed to division but the integers are not closed to division.

e) For the integers, the operation $*$ (where $a*b$ means 'multiply a by b^2') is not distributive over addition nor over multiplication.

3   A settee costs twice as much as an armchair. If they both go up by £150, a settee will cost $1\frac{1}{2}$ times as much as an armchair. How much does each cost? (*Hint* Let the armchair cost £x and the settee £$2x$.)

4   a) On graph paper draw the region defined by the orderings $y < x+3$, $y+x > 5$ and $x < 6$.

b) Give the co-ordinates of the vertices of this region.

c) Are the following points inside or outside the region:
   (2, 2)   (4, 4)   (5, 7)?

✱   5   a) Give the value of *i*) $4 \div 5$ (mod 11), *ii*) $3 \div 6$ (mod 7).

b) State which numbers have square roots in the arithmetics to modulus 11.

c) On the Farey lattice, name three points on each of the lines representing
   i) $\frac{3}{5}$ *ii*) $\frac{5}{11}$ *iii*) $\sqrt{2}$

11 Percentages

11A

1 Write the following percentages as fractions in their simplest form:

25%   40%   16%   85%   120%   36%   12½%   64%

33⅓%   8⅓%   18¾%   200%   37·5%   13·75%   72·4%

2 Using some of the fractions you have just found, work out the following:

a) 25% of £160 b) 40% of £5 c) 16% of £625
d) 85% of £600 e) 120% of £500 f) 36% of £125
g) 12½% of £96 · h) 8⅓% of £360 i) 13·75% of £800
j) 72·4% of £750

3 Find the value of each of the following:

a) 65% of £44 b) 48% of 5 kg c) 7½% of 200 litres
d) 57⅓% of 15 litres e) 26·4% of 50 km f.) 112% of £55
g) 6·25% of 108 km h) 24% of 1 km i) 145% of 1 hour
j) 8·75% of 52 kg

4 An oil tank which holds 920 litres is only 35% full. How much oil is there in the tank?

5 How much water is needed to fill a 240 litre water tank which is only 15% full?

6 In a school of 625 pupils, 52·8% are boys. a) How many boys are there? b) What percentage are girls?

7 Out of 450 cars leaving a factory, 17⅓% are yellow. How many are yellow?

8 A man invested £1240 in a financial trust, paying interest at the rate of 8¾% per year. What would his interest amount to after one year?

9 A grocer bought goods worth £960 from a wholesaler. If he reckoned to make 12% profit, how much profit should he expect on this consignment?

10 In an examination which is part theory and part practical work, the total marks awarded are 275. If 88% of the marks are for theory, how many marks are given for practical work?

11B

1 In each of the following, express the first quantity as a percentage of the second:

a) 25p, 50p b) 7 m, 25 m c) 14 cm, 40 cm

d) 36 m, 75 m	e) 22 kg, 80 kg	f) 120 m, 80 m
g) 27 p, 72 p	h) £1.98, £1.44	i) 35 cm, 1 m
j) 64 cm, 1·25 m	k) 273 g, 1·95 kg	l) 5 mins, 1 hour
m) 198 cm³, 3·6 litres	n) 798 g, 1·68 kg	o) £1.61, 46p

2   Out of the 500 seats in a theatre only 382 are filled. What percentage are full?

3   At the end of a term, 108 out of a total of 675 books in a school library have not been returned. What percentage of the total number of books are actually there?

4   A jar of jam which cost 40p had its price increased to $42\frac{1}{2}$p. What is the actual increase in price? What is this increase expressed as a percentage of the original price?

5   The area of Mr Green's garden is 540 m². He estimates that he has 300 m² of lawn. What percentage of his garden is lawn? Give your answer correct to 1 decimal place.

6   A man earning £75 a week manages to save £12 a week. What percentage is he spending?

7   20 pupils in a class of 32 stay at school for dinner. What percentage are staying?

8   The area of the walls of a room is 46·2 m². Of this area the door and windows occupy 6·6 m². What percentage of the walls are either door or windows? (Give your answer correct to 1 decimal place.)

9   40 of the residents in Church Street had television sets, and 24 of these were colour sets. What percentage were black and white?

10   The railway network in a rural area was examined to see which sections were making heavy losses and which sections were no longer useful to the public. It was decided to close 65 km out of a total of 1250 km. What percentage was to be closed?

11C Cost Price and Selling Price

Note Profit is reckoned as a percentage of the cost price unless otherwise stated.

1   With the given information, find the selling price of each of the following:

a) Cost 50p, profit 10 %   b) Cost £2.40, profit 15 %
c) Cost £16, profit $7\frac{1}{2}$ %   d) Cost £38, loss 22 %
e) Cost £115, loss 20 %   f) Cost £2500, profit 12·4 %
g) Cost £7.25, loss 8 %

2 In each of the following, find the loss or gain expressed as a percentage of the original cost:

a) Cost £3.50, selling price £3.78

b) Cost £145, selling price £159.50

c) Cost 65p, selling price 78p

d) Cost £1.20, selling price 96p

e) Cost £4600, selling price £4761

f) Cost £10.75, selling price £9.46

g) Cost £64.40, selling price £56.35

3 a) What is the selling price of an article bought for £50 and sold to gain 10%?

b) Express the selling price as a percentage of the cost price.

4 a) What is the selling price of an article bought for £60 and sold at a loss of 10%?

b) Express the selling price as a percentage of the cost price.

5 In each of the following, find the original cost:

a) Profit 5%, selling price £105

b) Profit 10%, selling price 44p

c) Profit 12%, selling price £224

d) Loss 6%, selling price £47

e) Loss $2\frac{1}{2}$%, selling price 39p

f) Profit $7\frac{1}{2}$%, selling price £387

g) Profit 24%, selling price £74.40

6 A house bought for £6500 some years ago has just been sold for £13 988. What percentage profit does this represent?

7 A three-piece suite is marked down to £294 in a sale with a notice to say that this is a saving of 16% of its normal price. What was its usual price?

8 A rocking horse is usually sold for £30.40, but is reduced by 40% in a sale.

How much is saved and what is its reduced price?

9 A gas heater is reduced by 27% of its recommended price to sell at £51.10. What was its recommended price?

10 A different type of heater with a recommended price of £72 is in the sale at £60. By what percentage has the usual retail price been reduced? (Give your answer correct to 1 decimal place.)

11 In a sale, a jacket is reduced by 40% of its marked price, and is sold for £12. What was the marked price?

12 A tin of minced beef is advertised at 27p instead of its usual price of 40p. Express the reduction as a percentage of its usual price.

11D Miscellaneous

1 In a sale, a child's cycle said to be worth £20.40 is reduced to £15.98. What is the actual reduction? Express this saving as a percentage of the usual price. (Give your answer correct to 1 decimal place.)

2 A firm which produces sectional garages is selling two types cheaply in a sale. Type *A*, usually £330, is reduced to £180. Type *B*, usually £375, is reduced to £200. Find the actual saving on each type, and in each case express this saving as a percentage of the usual price. Which is the better buy? (Give answers correct to 1 decimal place where necessary.)

3 The adult fare for a day trip is £2.10. A child's fare is $66\frac{2}{3}\%$ of the adult's. How much would it cost for a child? How much would it cost for mother, father and two children?

4 On another day trip a child's fare is £3.50 and an adult's £4.50. What percentage of the adult's fare is the child's? (Give your answer correct to 1 decimal place.)

5 An oil tank contains 390 litres and is 30% full. How much does the tank hold when full?

6 A garden fertiliser is said to consist of three main ingredients: nitrogen 16%, phosphoric acid 21%, potash 27%.
  a) What percentage is made up of other ingredients?
  b) In a bag containing 2.75 kg of the fertiliser, what weight of nitrogen is there?

7 During a strike a factory lost output worth £3 000 000. If this represents 48% of their monthly target, how much should they have produced in a month?

8 $8\frac{1}{3}\%$ of the area of a small farm of total area 40·5 hectares is used for growing turnips. How many hectares of turnips are there?

9 A farmer increases the size of his farm to 179·2 hectares by buying more land, equivalent to 12% of his original holding. What was the size of his farm to start with?

10 A greengrocer opens three boxes each containing 48 peaches. In the first box 5 are bad, in the second only 1 is bad, and in the third 3 peaches are bad. What percentage of the contents of his three boxes is bad?

11 A union are asking for a pay increase of 22% for their members, but in the end only get a $12\frac{1}{2}\%$ rise. For a man earning £4250 a year, what is the difference between the increase he hoped to get and the one he actually gets?

12 A report put out by building societies stated that on an average new owner-occupiers were paying £10 800 for a house, and of this 77·5% was borrowed from a society. If this is the case, what is the average amount of money being paid by the first-time buyer from his own resources?

13 A company announced that their profits of £30·6 m showed an increase of 70% over the previous year's. What were their profits in the previous year?

14 A car bought for £3 600 is sold some months later at a loss of 26%.
a) How much was it sold for?
b) If the second purchaser sells later, losing 25% of the price he paid, what does he sell for?
c) What is the difference between this final selling price and its original cost? Express this difference as a percentage of the original purchase price.

Express this difference as a percentage of the original.

15 A rectangle measures 10 cm by 8 cm. Its length and breadth are both increased by 25%. Find the percentage increase in its area.

16 A dealer made a profit of $12\frac{1}{2}$% by selling a horse for £216. What did he pay for it originally? Find the percentage profit that he would have made if he had sold the horse for £240.

17 An engineer whose salary is £4000 has a rise of 6%. Some time later salaries are reduced by 5%. Is he better or worse off than he was originally? What is the difference between his original and final salaries?

18 A rectangular solid wooden block measures 5 cm by 4 cm by 3 cm. Wood is cut away until only 45% of the original volume remains and the wood is then in the shape of a cube. Find the length of the edges of the cube.

19 A doctor bought a car for £3000. In the first year its value decreased by 15%. In the second year it decreased by 12% of what it was at the beginning of that year. In the third year it depreciated by 10% of its value at the start of that year. How much was the car worth after 3 years?

20 In a local election, 1200 people in one village were entitled to vote. Only 65% of them voted, and of these 40% were women. How many men actually voted?

12 More about Similarity

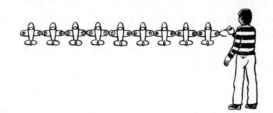

12A Plane Figures
(i.e. Two-Dimensional)

1   a) Draw a square of side 3 cm. What is its area?
    b) Draw a square of side 6 cm. What is its area?
    c) Draw a square of side 9 cm. What is its area?
    d) What is the ratio of the lengths of the sides of the three squares?
    e) What is the ratio of the areas of the three squares?
    f) What do you notice about the two ratios in d) and e)?

2   a) Draw a rectangle of sides 3 cm and 4 cm. What is its area?
    b) Repeat question a) with sides 6 cm and 8 cm.
    c) Repeat question a) with sides 9 cm and 12 cm.
    d) Write down the ratio of the lengths of the sides of the three rectangles.
    e) Write down the ratio of the areas of the three rectangles.
    f) What do you notice about the two ratios in d) and e)?

3   a) Draw a triangle with base 4 cm and height 5 cm. What is its area?
    b) Draw another triangle with base 6 cm and height 7·5 cm.
    What is its area?
    c) Draw a third triangle with base 8 cm and height 10 cm.
    What is its area?
    d) Express the ratio of the lengths of the base of the three triangles in the
    simplest possible way, using whole numbers.
    e) Repeat d) for the heights.
    f) Repeat d) for the areas.
    g) Comparing the ratios in d), e) and f) what do you notice?

4   a) Draw a parallelogram with two parallel sides of length 4·8 cm, the
    distance between them being 3·6 cm. What is its area?
    b) Draw another parallelogram with parallel sides of length 9·6 cm,
    the distance between them being 7·2 cm. What is its area?
    c) What is the ratio of the lengths of the sides of the two parallelograms?
    d) What is the ratio of the areas of the two parallelograms?
    e) What do you notice about the two ratios in c) and d)?

5   a) Draw a trapezium with parallel sides of 2 and 4 cm respectively, the
    distance between them being 3 cm. What is its area?

95

b) Draw another trapezium in which all the dimensions in *a*) are increased by 50%. Find its area.

c) Draw a third trapezium in which all the dimensions in *a*) are doubled. Find its area.

d) What is the ratio of the lengths of corresponding sides in these three trapeziums? Put this ratio in its simplest possible form using whole numbers only.

e) Repeat *d*) for the areas of the three trapeziums.

f) What do you notice about the ratios in *d*) and *e*)?

6 *a*) Draw a circle of radius 2 cm. By stepping off lengths equal to the radius in the usual way, mark the six vertices of a regular hexagon. Join these up to form the hexagon, and join each vertex to the centre. Measure the height of one of the six triangles which have their vertices at the centre. Hence calculate the area of one triangle. What is the area of the hexagon?

b) Repeat *a*) with a circle of radius 3 cm.

c) Repeat *a*) with a circle of radius 5 cm.

d) Write down the ratio of the lengths of the sides of the three hexagons in the form $1 : a : b$.

e) Write down the ratio of the areas of the three hexagons, in the form $1 : c : d$.

f) Allowing for drawing errors, what do you notice about the two ratios in *d*) and *e*)?

7 On a sheet of graph paper plot the following points, using 2 cm as the unit, and the *x*-axis parallel to the longer side of the paper.

a) (1, 3)   (3, 5)   (6, 2)   (2, 1)

b) (2, 6)   (6, 10)   (12, 4)   (4, 2)

c) (3, 9)   (9, 15)   (18, 6)   (6, 3)

d) In each case measure the length of the longest side as accurately as you can.

e) By 'boxing in', or otherwise, find the area of each figure.

f) Write down the ratio of the lengths of the longest sides for each of the three figures.

g) Write down the ratio of the areas of the three figures.

h) What do you notice about the ratios in *f*) and *g*)?

8 Repeat question 7, with co-ordinates

a) (2, 1)   (5, 3)   (4, 4)   (2, 4)   (1, 2)

b) (4, 2)   (10, 6)   (8, 8)   (4, 8)   (2, 4)

c) (3, 0)   (12, 6)   (9, 9)   (3, 9)   (0, 3)

9 Repeat question 7 with co-ordinates

a) (1, 1)   (5, 2)   (4, 4)   (2, 3)

b) (2, 2)   (10, 4)   (8, 8)   (4, 6)

c) (0, 2)   (12, 5)   (9, 11)   (3, 8)

10 The sets of figures in each of questions 1 to 9 were similar. If you answered correctly, you will now know that the areas of similar figures are proportional to the squares of the lengths of corresponding sides. This is true whatever the shape of the figure.

a) If the area of an irregular octagon, one of whose sides is 7 cm long, is 147 cm² , what would be the area of a similar octagon in which the corresponding side was 9 cm?
b) If the area of a kite was 28·5 cm² , what would be the area of a similar kite whose sides were all 10 % longer?
c) Repeat question b) for a kite whose sides are 10 % shorter.
d) If an irregular figure is drawn on graph paper and then enlarged by a scale factor of 1·5, what will be the effect on its area?
e) Repeat question d) with an enlargement by a scale factor of ½.

11 a) In two similar figures, the ratio of the lengths of the sides is 1 : 3. What is the ratio of the areas?
b) Repeat a) with the sides in ratio 2 : 5.
c) Repeat a) with the sides in ratio 3 : 7.
d) Repeat a) with the sides in ratio m : n.

12 If three similar figures have areas in the ratio p : q : r, what is the ratio of the lengths of corresponding sides?

12B Solid Figures (i.e. Three-Dimensional)

1 a) Think of a cube of side 4 cm, made up of little cubes, each of side 1 cm.
    i) How many little cubes will there be in the bottom layer of the big cube?
    ii) How many layers will there be?
    iii) How many little cubes will there be altogether?
    iv) What is the volume of one little cube?
    v) What is the volume of the big cube?
  b) i) What is the area of the base of the big cube?
     ii) What is the height of the big cube?
  c) In view of your answers to a) and b), is it true to say that the volume of a cube is 'area of base x height'?

2 a) What is the volume of a cube of side 2 cm?
  b) What is the volume of a cube of side 3 cm?
  c) What is the volume of a cube of side 5 cm?
  d) What is the ratio of the lengths of the sides of these three cubes? (written in its simplest form, using whole numbers only)
  e) What is the ratio of the volume of these three cubes? (written in its simplest form, using whole numbers only)
  f) What do you notice about the two ratios in d) and e)?

3 a) What is the volume of a rectangular prism (or a cuboid, or a 'box') whose sides are of lengths 3, 4 and 5 cm respectively?

97

b) If it stands on its smallest side (3 cm × 4 cm) what is the area of its base? What is its height? What is the product 'area of base × height'? Is this the same as the volume you calculated in *a*)?

c) Repeat *b*) for the prism standing on the side 3 cm × 5 cm.

d) Repeat *b*) for the prism standing on the side 4 cm × 5 cm.

e) Is it true to say that the volume of the prism is 'area of base × height' whichever side is taken as area of base?

f) Is it true to say for *any* prism that the volume is 'area of base × height'?

4 *a*) What is the volume of a rectangular prism whose sides are 2 cm, 4 cm and 7 cm?

b) Repeat question *a*) with sides 4, 8, and 14 cm.

c) Repeat question *a*) with sides 6, 12, and 21 cm.

d) What is the ratio of the lengths of corresponding sides in the three prisms? Write this ratio in its simplest form using whole numbers only.

e) Repeat *d*) for the volumes of the three prisms.

f) What do you notice about the ratios in *d*) and *e*)?

5 *a*) The base of a prism is a pentagon of area 20 cm². Its faces are all perpendicular to the base (i.e. it is a right prism). Its height is 6 cm. What is its volume?

b) A similar prism has a height of 12 cm.

i) What is the area of its base? *ii*) What is its volume?

c) Another similar prism has a height of 24 cm.

i) What is the area of its base? *ii*) What is its volume?

d) What is the ratio of the heights of the three prisms expressed in its simplest form using whole numbers only?

e) What is the ratio of the areas of the bases of the three prisms expressed in its simplest form using whole numbers only?

f) What is the ratio of the volumes of the three prisms, expressed in its simplest form, using whole numbers only?

g) What do you notice about the three ratios in *d*, *e*) and *f*)?

6 *a*) How many faces are there to a cube? What is the total area of the faces of the cube in question 1?

b) What is the total area of the faces of each of the 3 cubes in question 2?

c) Write down the ratio of the three areas in *b*) in its simplest possible form, using whole numbers only.

d) Write down also the ratio of the lengths of the sides of the 3 cubes in *b*).

e) What do you notice about the two ratios in *c*) and *d*)?

7 *a*) What is the total area of the sides of each of the three rectangular prisms in question 4?

b) Express the ratio of these three areas in the simplest possible way, using whole numbers only.

c) Write down the ratio of the lengths of corresponding sides in these three prisms, using whole numbers only and expressing the ratio in its simplest form.

d) What do you notice about the ratios in *b*) and *c*)?

✳ *8* *a*) The area of the base of a cylinder is 48 cm² and its height is 10 cm. What is its volume?
b) Another similar cylinder is 25 % taller. What is the area of its base? What is its volume?
c) A third similar cylinder is 75 % taller. What is the area of its base? What is its volume?

THE VOLUME OF A PYRAMID IS ⅓ OF THE AREA OF BASE × HEIGHT.

✳ *9* *a*) Two similar pyramids have bases of area 20 cm² and 80 cm² respectively. *i*) What is the ratio of these areas, expressed in its simplest form? *ii*) What is the ratio of their heights?
b) What is the ratio of their volumes?
c) If the height of the first is 15 cm, what is the height of the second?
d) You will learn in a later book that the volume of a pyramid is ⅓ of the area of base × height. What are the volumes of the two pyramids?
e) Does your answer in *d*) agree with the ratio of volumes you calculated in *b*)?

✳ *10* Each of the faces of a regular tetrahedron has an area of 4 cm².
Each of the faces of another regular tetrahedron has an area of 9 cm². What is the ratio of the lengths of their sides? What is the ratio of their volumes?

✳ *11* The slant length of a cone, i.e. the distance measured from the vertex to any point on the circumference of the base, is 6 cm and the total surface area (i.e. the area of the slant surface and the base) is 36 cm². If the slant length of a similar cone is 8 cm, what is the total surface area of this second cone?

✳ *12* A chemical compound which forms crystals in the shape of regular octahedra is being allowed to crystallise out from aqueous solution. If the distances between opposite vertices of two crystals are 3 mm and 5 mm, what is the ratio of the weights of these two crystals?

✳ *13* A class is making models of a stellated dodecahedron. First they make a regular dodecahedron, and then on each pentagonal face they stick a pyramid of the right height. Peter and Philip are using pyramids of height 8 cm. Joan and Susan are using pyramids of height 10 cm. Express the area of card Peter and Philip use altogether as a percentage of the area of card Joan and Susan use.

✳ *14* Mr and Mrs Scott are taking the children on holiday. They buy a beach ring of 24 cm diameter for Ann, aged 4, and another of 36 cm diameter for John, aged 8.

a) What is the ratio of the surface areas of these two rings?
b) What is the ratio of their volumes?
c) If it takes 32 strokes of the pump to inflate Ann's ring, how many strokes would it take to inflate John's?

1 A firm markets 'Staysweet' bath salts in three sizes of container. All the containers are shaped like bears, and all three sizes are similar to one another. If the heights of the bears are 10 cm, 12 cm and 15 cm respectively, and the smaller size sells at 32p, what would be a fair price for each of the others?

2 Ornamental glass balls are packed for transit in cardboard boxes which are cubical, the inside being filled with packing, and the diameter of each ball being three quarters of the length of the side of the cube. One face of each cube is marked with red lines along the diagonals. If the length of the diagonal is 20 cm for one size of ball and 30 cm for another size of ball:

   a) What is the ratio of the diameters of these balls?
   b) What is the ratio of their surface areas?
   c) What is the ratio of their volumes?

3 A bottle of orange squash stands 32 cm high and contains enough to make 25 glasses of suitably diluted orange drink. A bottle of similar shape stands 24 cm high. *a)* What is the ratio of the volumes of these two bottles? *b)* How many glasses of orange drink would you expect to get from the smaller bottle?

4 So far in this chapter we have studied figures that are *similar*, i.e. if one dimension is increased in a certain ratio, every other dimension is increased in the same ratio. Here are some problems where the figures are not similar.

   a) A square bottle of side 8 cm holds 500 ml (i.e. half a litre). How much would another square bottle hold if it was the same height but its side was 10 cm?
   b) A square bottle whose height is 18 cm holds $\frac{3}{4}$ litre. How much would a bottle of the same cross section hold if it was 12 cm high?
   c) The area of the inside of the base of a cylindrical bottle is 12 cm² and it holds 100 ml. The area of the base of another bottle is 15 cm² and it is half as high again. How much does it hold?

5 A watering can holds 5 litres. Its bottom is an oval whose greatest length is 24 cm. In an exactly similar can the corresponding length is 30 cm. How much does the larger can hold? (Give a sensible approximate answer.)

6 A packet of sausages weighing $\frac{1}{4}$ kilogram contains eight sausages of standard length and diameter. How many sausages would there be in a $\frac{1}{4}$ kilogram packet if they were exactly the same shape but 10 % longer? Give a sensible answer (i.e. no fractions).

7 If a jar of anchovy paste has an internal diameter of 3 cm and costs 20p, what would be the fair price of an exactly similar jar of internal diameter 5 cm?

8 A firm makes three sizes of paddling pools, all in heavy plastic material. For toddlers, the pool is circular with an inside diameter of 2 metres. For seven-year-olds it has a diameter of 3 metres, and for ten-year-olds 4 metres. All other dimensions go up in the same ratio.

  a) If the smaller size holds 1000 litres of water, what do the other two pools hold?
  b) If the cost is proportional to the area of plastic, and the largest pool costs £100, what would be the cost of the other two pools? (Answer to the nearest pound.)

9 A woman makes her own doughnuts. From 350 g of strong white flour (with added sugar, milk, fat and yeast) she gets 16 doughnuts, each approximately spherical and of diameter 7 cm. She wants more doughnuts, so she reduces the diameter to approximately 6 cm. *a*) How many doughnuts will she now get? *b*) She sprinkles caster sugar on the outside. Assuming the weight of caster sugar is proportional to the surface area, will she use more or less caster sugar when she makes the smaller doughnuts?

10 *a*) Assuming that nails are of similar shapes (i.e. that the diameter is proportional to the length), write down a ratio that expresses the volumes of nails of the following lengths: 2·5 cm, 7·5 cm and 15 cm.
  b) If one kilogram of 15 cm nails contains 10 nails, how many would you expect in 1 kilogram of 2·5 cm nails? (Your answer will be unrealistic as the diameter is not proportional to the length.)

13 Simple Trigonometry

In this chapter either 3-figure or 4-figure tables can be used.
When using 4-figure tables for cosines, remember to subtract *the differences.*

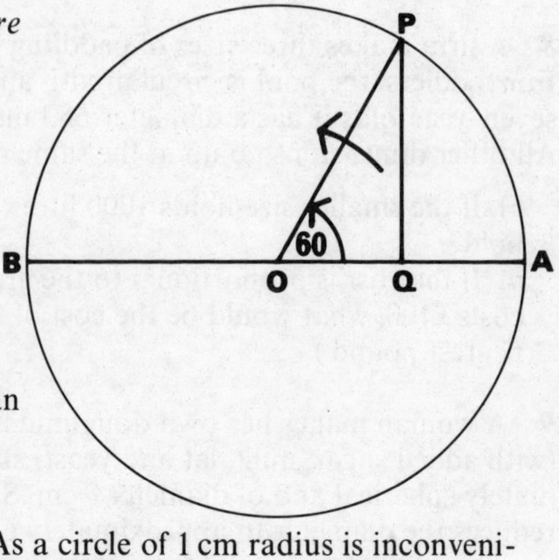

13A Sine and Cosine

1 The diagram shows a circle, centre *O*, and a diameter *AOB*.

A vector *OP* rotates anti-clockwise starting from *OA*, and has traced out an angle of 60°. A perpendicular is dropped from *P* to the diameter *AOB* meeting it at *Q*.

The circle should be of unit radius. As a circle of 1 cm radius is inconveniently small, a unit of 3 cm is used. By actual measurement you will find that *PQ* is 2·60 cm and *OQ* is 1·5 cm. Dividing by 3 gives *PQ* as 0·86 units and *OQ* as 0·5 units.

If the circle is drawn very accurately and on a larger scale (the radius still being taken as one unit) you will find that *PQ* was 0·866 units and *OQ* 0·500 units.

Now look up sin 60° and cosine 60° in the tables. What are the values? Does this mean that (provided the circle has unit radius) the length of *PQ* is the sine of the angle at *O* and the length of *OQ* is the cosine of the angle at *O*.

2 Draw the diagram in question 1 as accurately as you can, using 5 cm or 10 cm as a unit (you need only draw one quarter of the circle). Measure *PQ* and *OQ*. *PQ* should be 5 × sin 60° (or 10 × sin 60°) and *OQ* should be 5 × cosine 60° (or 10 × cosine 60°). See how close you can get to these answers.

3 Repeat question 2 with angles of 15°, 30°, 50° and 70°. In each case measure the lengths of *PQ* and *OQ* and draw up a table as follows:

1	2	3	4	5	6	7	8
Length of *OP*	Angle at *O*	Length of *PQ*	Length of *OQ*	*PQ* ÷ 5	*OQ* ÷ 5	Sine of angle at *O*	Cosine of angle at *O*
5 cm	15°						
5 cm	30°						
5 cm	50°						
5 cm	70°						

The table is drawn up for a unit of 5 cm. If a unit of 10 cm is used, the table must be modified accordingly.
Columns 3 and 4 are to be filled in by actual measurement.
Columns 7 and 8 should be filled in using the tables.
Columns 5 and 6 should agree with columns 7 and 8 reasonably well.
See how close you can get.

4 Use your tables of sines and cosines to find the lengths of $a, b, c, d, e \ldots$ in the following triangles (not drawn to scale).

Example If the radius vector had been 1, a would have been sin 60°; i.e. 0·866. But the radius vector is 3, so a is $3 \times 0\cdot866$, i.e. 2·60 (to 3 s.f.).

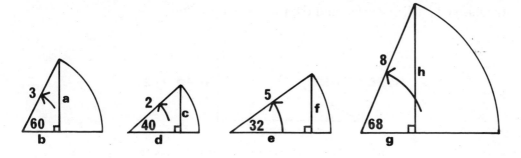

5 Repeat question 4 with the following triangles. (The arc of the circle is no longer shown.)

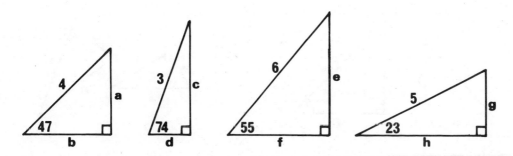

6 In each of the following triangles you are given two lengths. Find the angles $A, B \ldots F$. (The triangles are not drawn to scale.)

Example Cos $A = \frac{3}{5} = 0\cdot600$ $A = 53\cdot1°$ (using 3-figure tables) or 53°8′ (using 4-figure tables).

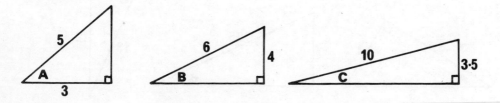

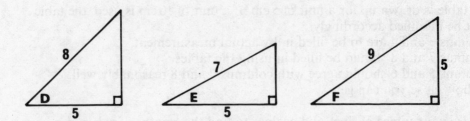

7 From this point on we shall think about right-angled triangles with the right angle in any position. We shall find it helpful to call the side opposite the right angle the *hypotenuse*, the side which (with the hypotenuse) encloses the marked angle *adjacent*, and the third side *opposite*.

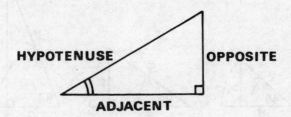

Pick out the hypotenuse and the opposite and adjacent sides in the following triangles.

Example Hypotenuse *a*, adjacent *c*, opposite *b*.

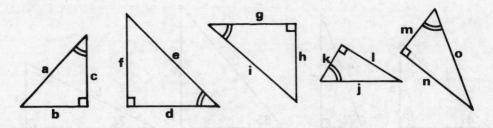

8 We can now say

$$\sin A = \frac{\text{Length of opposite side}}{\text{Length of hypotenuse}} \quad \text{or more simply} \quad \frac{\text{opposite}}{\text{hypotenuse}}.$$

State the values of sin *A*, sin *B* etc. in the following triangles (not drawn to scale). Leave your answers as fractions.

Example Sin $A = \frac{4}{5}$

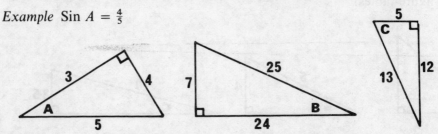

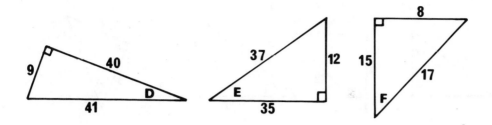

9 We can also say that

$$\text{cosine } A = \frac{\text{length of adjacent side}}{\text{length of hypotenuse}} \quad \text{or more simply} \quad \frac{\text{adjacent}}{\text{hypotenuse}}.$$

Using the triangles in question 8, write down the values of cos *A*, cos *B* etc. leaving your answers as fractions.

10 Find the angles *L, M ... Q* in the following triangles (not drawn to scale).

Example Cos $L = \frac{5}{7.5} = 0.6667$ $L = 48.2°$ (using 3-figure tables) or 48°11′ (using 4-figure tables).

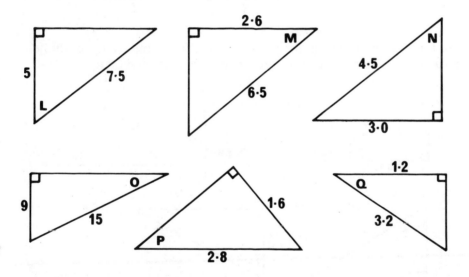

11 Calculate the lengths *a, b ... j* in the following diagrams (not drawn to scale).

Example $a = 2.5 \times \cos 42° = 2.5 \times 0.743 = 1.86$ (to 3 s.f.).

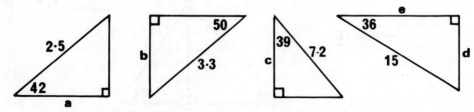

105

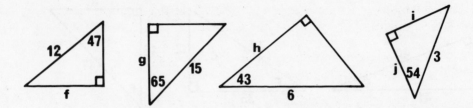

12 Use tables to find the length x in the first diagram. Then find the lengths a and b in the other two diagrams. What can you say about these three triangles?

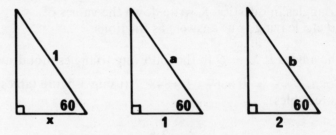

13 Use tables to find the lengths $p, q \ldots$ in the following triangles. (Note that in each case you are asked to find the length of the hypotenuse.)

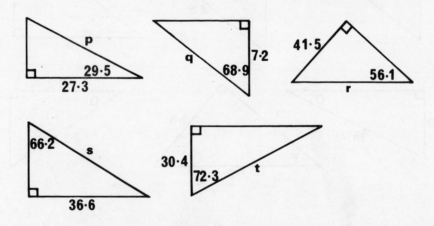

* **14** If $AB = 6$ cm, find the length of DB.

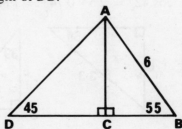

15   Find the lengths of the sides of a rectangle whose diagonals are 12·5 cm long and make angles of 28° with the longer sides.

16   Find the angles of an isosceles triangle *ABC* in which *AB* = *AC* = 6·3 cm and *BC* = 9·8 cm.

17   A ladder 3·5 m long has to reach a window 3 m up a wall. Calculate the inclination of the ladder to the ground, and find how far the foot of the ladder must be from the foot of the wall.

13B

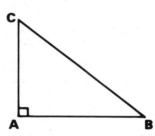

1   *a)* If *BC* = 4 cm, ∠ *B* = 62° find *AB*
   b) If *BC* = 9 cm, ∠ *C* = 48·5° find *AB*
   c) If *BC* = 2·7 cm, *AB* = 0·9 cm find ∠ *B*
   d) If *BC* = 8·5 cm, *AC* = 3·4 cm find ∠ *B*
   e) If *AC* = 6 cm, ∠ *B* = 30° find *BC*

You will find that it helps you if you draw a separate triangle for each part of the question and put in the measurements given.

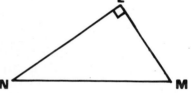

2   *a)* If *MN* = 16 cm, ∠ *N* = 25° find *LN*
   b) If *MN* = 20 cm, ∠ *M* = 34° find *LN*
   c) If *MN* = 66 cm, *LN* = 36 cm find ∠ *M*
   d) If ∠ *N* = 66°, *LM* = 4 cm find *MN*
   e) If *MN* = 42 cm, *LM* = 30 cm find ∠ *M*

Once again it will help you to draw five different triangles and put in the different measurements given.

3   Find the lengths *a, b, c, d, e* in the following triangles:

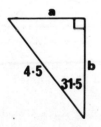
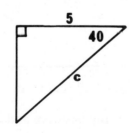
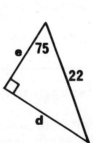

4   Find the angles *F, G* ... in these triangles:

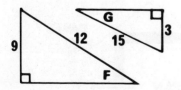
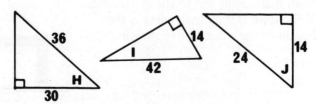

107

5 Find the length of the diagonals of a square of side 10 cm.

6 A boy is flying a kite on a string 16 m long. When the inclination of the string to the horizontal is 70°, how high is the kite? (Assume the string remains straight.)

7 The spoke of a wheel is 0·9 m long. The wheel is standing on the ground and the spoke makes an angle of 18° with the upward vertical. What is the height of the upper end of the spoke above the ground? (Ignore the thickness of the rim.)

8 A ship sails 25 km on a bearing of 068° from a port. What is its distance *a*) due north and *b*) due east of the port?

* **9** The diagram shows a regular pentagon of side 5 cm. Find the distance of D from AB.
(*Hint* Find angles EAD and DAB.)

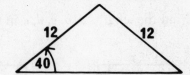

* **10** Find the lengths of the diagonals of a rhombus of side 12 cm if two opposite angles are each 120°.

11 The sides of an isosceles triangle are 12 cm and the base angles are 40°.
a) What is the length of the base? *b*) What is the height? *c*) What is the area?

(*Hint* Remember that the perpendicular from the vertex to the base bisects the base and also bisects the vertical angle.)

12 *a*) If the base of an isosceles triangle is 16 cm, and the base angles are 50°, what is the length of the sides? *b*) What is the height of the triangle?

13 The height of an isosceles triangle is 8 cm and the sloping sides are each 11 cm. *a*) What are the base angles, and *b*) what is the length of the base?

14 The two sides of a sloping roof are both inclined at 38° to the horizontal. If the span is 9 metres, what is the height of the ridge above the eaves?

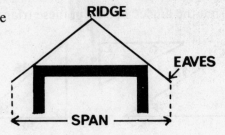

15 The roof of a small factory has a span of 12 metres. One side of the roof faces south and is sloped at an angle of 30° to the horizontal. The other side faces north and is sloped at an angle of 60° to the horizontal.
a) What are the lengths of these sloping sides?
b) How high is the ridge above the eaves?

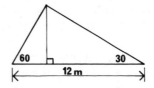

16 Triangle *ABC* has base *BC* 7·5 cm and side *AB* 6 cm. If angle *B* is 52°, calculate the height of the triangle and its area.

17 A parallelogram has sides of length 14 cm and 11 cm respectively. If the angle between a pair of sides is 67°, find the distance between the longer sides, the distance between the shorter sides, and the area. (*Note* 'Distance' means 'perpendicular distance'.)

18 Another parallelogram has an area of 60 sq cm, and its sides are of length 8 cm and 10 cm. What is the angle between the sides?

19 A triangle has an area of 37 cm², a base of length 10·5 cm and a side of length 8·4 cm. What is the angle between this side and the base?

∗ 20 A path slopes steadily up the face of a straight vertical cliff at an angle of 8° for a distance of 83 metres. At this point a shelter has been constructed in the cliff, and seats are available. The path then slopes back in the opposite direction at an angle of 9°, for a further 92 metres. Here there is a second shelter: a large flat area has been excavated, telescopes have been installed for viewing the shipping in the sea lanes below, and tea, coffee, ice cream, soft drinks, etc. can be purchased. A flight of 36 steps then takes you to the top of the cliff, each step being 20 cm high.
   Calculate *a)* the height of the first shelter above the foot of the cliff, *b)* the height of the second shelter and *c)* the total height of the cliff (all answers to the nearest metre).
   What parts of the above information are superfluous?

∗ 21 An aeroplane flies 93 kilometres on a bearing of 056°. It then changes course and flies 211 km on a bearing 158°. *a)* How far is it east from its starting point? *b)* How far is it south of its starting point? Give your answers to the nearest kilometre.
(*Hint* Draw a sketch of the whole flight, and separate sketches of each leg of the flight.)

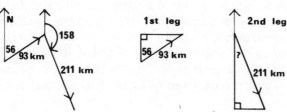

* **22** A submarine cruises 32 km on a bearing of 067° from its base and then receives instructions from its base to change course and sail on a bearing of 136° for a distance of 43 km to intercept a ship carrying contraband. It makes a perfect interception and the vessel is boarded. How far was the submarine east of its base at the moment of boarding, and how far north or south?

Give your answers to the nearest tenth of a kilometre.

* **23** A speedboat leaves its port with a full complement of passengers and steers a straight course to a flashing beacon which is moored 6·3 km north-west of the port. From the beacon it speeds on a straight course on a bearing of 217° to a lightship which is moored over a dangerous sandbank, 9·1 km from the flashing beacon. It sails round the lightship and returns to port. How far is the lightship *a)* north or south *b)* west of the port?

Give your answers to the nearest tenth of a kilometre.

13C Tangents

Note The geometrical meaning of a tangent is a straight line that touches a circle in one point only and does not cut through it. By symmetry it must be at right angles to the diameter at the point where it touches the circle.

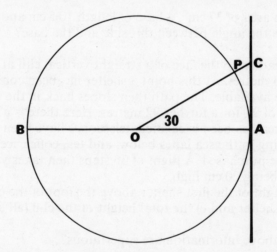

1 The diagram shows a circle, centre *O*, and a diameter *AOB*. It is very like the diagram in 13A, but the rotating vector *OP* this time has swept out an angle of 30°, and instead of the perpendicular *PQ* being drawn, the tangent is constructed at *A*, and the radius vector *OP* extended to cut this tangent in *C*. Once again the circle should have unit radius, but for convenience a radius of 3 cm has been used, so all measured lengths must be divided by 3.

Actual measurement will show that *AC* is 1·75 cm. Dividing by 3 we get 0·58 cm. If a more accurate diagram were drawn with a larger radius, still taking the radius as one unit we should find that *AC* was 0·577. Looking at the tables, what is the tangent of 30°? Does this mean the length cut off on the tangent at *A* by the radius vector (extended) is equal to the tangent of the angle at *O*?

110

2 Draw the diagram above as accurately as you can, using a radius of 5 or 10 cm. You need only construct one quarter of a circle. Measure AC and divide it by 5 (or 10) and see how close your answer is to tan 30°, i.e. to 0·577.

3 Repeat question 2 with angles of 15°, 25°, 35° and 45°. In each case measure the length of AC and divide by 5 (or 10). Enter your results in a table as follows:

Length of OP	Angle at O	Length of AC	$AC \div 5$	Tangent of angle at O
5 cm	15°			
5 cm	25°			
5 cm	35°			
5 cm	45°			

This table is drawn up for a unit of 5 cm, so it must be modified if a unit of 10 cm is used.
Column 3 should be completed by actual measurement.
Column 5 should be completed from the tables.
Columns 4 and 5 should agree closely. See how close you can get.

4 From Triangle OAC in the figure of question 1 we see that the tangent of the angle at O can be written as $\dfrac{\text{length of } AC}{\text{length of } OA}$ or $\dfrac{\text{length of opposite side}}{\text{length of adjacent side}}$

or more simply $\dfrac{\text{opposite}}{\text{adjacent}}$.

Use your tables to find the values of A, B, C etc. in the following triangles which are not drawn to scale.

Example Tan $A = \frac{3}{4} = 0·75$
$A = 36·9°$ (using 3-figure tables) or $A = 36° 52$ (using 4-figure tables)

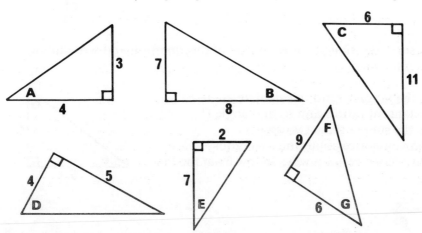

5 Use tables to find the lengths of a, b, c, d, e in the following triangles.

Example $\frac{a}{2} = \tan 40° = 0.839$

$\qquad a = 2 \times 0.839 = 1.68$ (to 2 d.p.)

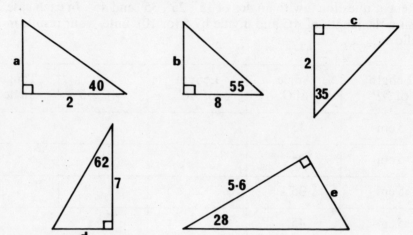

6 Find the lengths x and y.

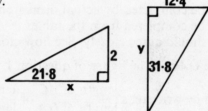

7 *a*) If $AB = 3.2\,\text{cm}$, $\angle A = 25°$ find BC
   b) If $BC = 7.5\,\text{cm}$, $\angle C = 39°$ find AB
   c) If $AB = 6\,\text{cm}$, $BC = 8\,\text{cm}$ find $\angle C$
   d) If $AB = 12\,\text{cm}$, $BC = 5\,\text{cm}$ find $\angle A$
   e) If $AB = 3.5\,\text{cm}$, $BC = 2\,\text{cm}$ find $\angle C$

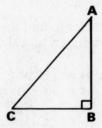

Draw a separate triangle for each part of this question and write in the given measurements.

8 Use the tangent ratio to calculate angle P.
Use the tangent ratio again to find angle Q.
What is the sum of these two angles?
What do you notice about the two ratios?
Does the second result always follow from the first?

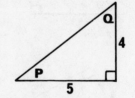

9 If you have answered question 8 correctly you will see that it is often simpler to work with the third angle of the triangle rather than with the given angle. Using this method where appropriate, calculate the lengths $k, m, n \ldots s$ in the following triangles.

Example To find k: the third angle (or complement) is $38°$.
$$\tfrac{k}{3} = \tan 38° = 0.7813$$
$$k = 3 \times 0.7813 = 2.34 \text{ (to 3 s.f.)}$$

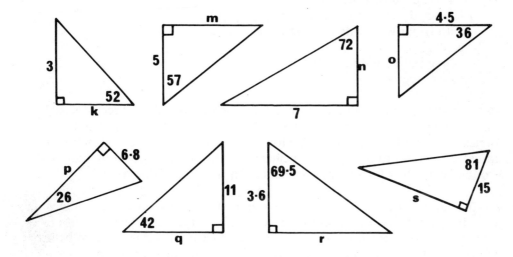

10 A rectangle measures 13 cm by 7 cm. Find the angle which the diagonal makes with the shorter sides.

11 In the diagram ABC and ACD are two right-angled triangles.
If $DC = 6$ cm, $\angle ADC = 42.5°$ and $\angle ABC = 65°$, find the length of AB.

12 A plank of wood is resting over the top of a vertical wall 1·5 m high and the foot of the plank is 0·6 m from the foot of the wall on horizontal ground. What is the angle which the plank is making with the ground?

13 A man who is 1·6 m tall stands on top of a cliff 53 m high and looks down at a boat out at sea. If the angle that he looks down through (the angle of depression) is 35°, how far is the boat from the foot of the cliff? (Give your answer to the nearest metre.)

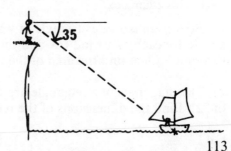

14 A ship sails due north for 36 km, then due east for 20 km. What is its final bearing from its starting point? Give your answer to the nearest degree.

✻ 15 *a*) An aeroplane on a training flight leaves the aerodrome and flies 53 km due west and then 18 km due south. What is its bearing from the aerodrome?
b) If it had flown the same distance due north instead of due south, what would its bearing have been?
     Give both answers to the nearest degree.

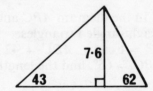

16 A radio mast standing on level ground is supported by three cables which are attached to a point 47 metres above the ground, and which make angles of 43° with the ground. The lower ends of the cables are attached by adjustable screws to heavy steel eyelets which are let into concrete blocks at ground level. How far are these eyelets from the foot of the mast?

17 A man sets out to row across a straight stretch of river to a landing stage directly opposite. At that point the river is 23 metres wide. The current is running strongly, however, and he is carried 17·1 metres downstream before he manages to land at some steps. His wife is standing at the point from which he started; she has a pair of binoculars which she originally focussed on the landing stage. Through what angle must she turn them to see her husband disembark? (Answer to nearest half degree.)

18 A triangle has a height of 7·6 cm and base angles of 43° and 62°.
What is the length of the base?
(*Hint* Use the complements of the given angles. See question 9.)

19 Repeat question 18 with a height of 9·7 cm and base angles of 37° and 71°.

✻ 20 Repeat question 18 with a height of 6·4 cm and sides of 8·5 cm and 10·2 cm. (*Hint* First find the two angles at the vertex.)

13D Miscellaneous

1 *PQR* is an isosceles triangle in which *PQ* = *PR* = 5 cm and angles *Q* and *R* are each 52°. Find *QR* and the perpendicular height of the triangle through *P*, then find the area of the triangle.

2 The diagonals of a rectangle are each 12 cm long and intersect at an angle of 72°. Find the dimensions of the rectangle.

114

3 A boy looks at a cross on a church spire through binoculars. He is 1·5 m tall and looks up through an angle of elevation of 68°. He estimates his distance from the foot of the spire to be 35 m. How high is the top of the spire? (Give your answer correct to the nearest tenth of a metre.)

4 *a*) Using this diagram write down the following ratios:
sin *C*, cos *B*, sin *B*, cos *C*.
What do you notice?
b) If sin 35° = 0·574 what is cos 55°?
If cos 62·5° = 0·462 what is sin 27·5°?
c) If sin 60° is 0·866 which other ratio can you write down?
d) If sin *X* = cos *Y* what do you know about angles *X* and *Y*?

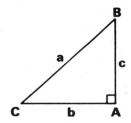

5 *PQRS* is a trapezium in which *PQ* is parallel to *SR*. *SR* = 2 m, *PQ* = 6 m, and the sloping sides make 45° with *PQ*. Find the area of the trapezium.

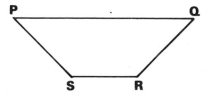

6 In Enderville the portion of High Street between the public library and the church has to be closed down for repairs and through traffic to Skintown is diverted via Furley Road and Cross Road. All these three roads may be taken as straight.

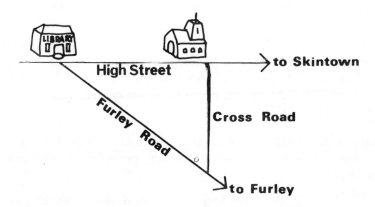

The length of High Street that is closed is 270 m. The approximate angle between High Street and Furley Road is 36°, and Cross Road meets High Street at right angles. How much further does a motorist travel in a single journey to Skintown because of the diversion? (Slide rule accuracy is sufficient.)

7 Calculate the acute angle between the following vectors and the axis stated.

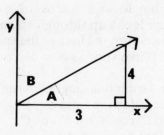

Example $\begin{pmatrix} 3 \\ 4 \end{pmatrix}$

Tan $A = \frac{4}{3} = 1{\cdot}3333 \quad A = 53{\cdot}1°$ or $53°8'$
Angle between vector and x-axis is A.
Angle between vector and y-axis is B, i.e. $90° - A$.

a) $\begin{pmatrix} 4 \\ 5 \end{pmatrix}$ x-axis b) $\begin{pmatrix} 7 \\ 4 \end{pmatrix}$ y-axis c) $\begin{pmatrix} 3 \\ 7 \end{pmatrix}$ y-axis d) $\begin{pmatrix} 6 \\ 2 \end{pmatrix}$ x-axis

e) $\begin{pmatrix} 5 \\ 5 \end{pmatrix}$ x-axis f) $\begin{pmatrix} 9 \\ 11 \end{pmatrix}$ x-axis g) $\begin{pmatrix} 1 \\ 4 \end{pmatrix}$ y-axis h) $\begin{pmatrix} 3 \\ 8 \end{pmatrix}$ y-axis

* 8 Repeat question 7 for the following vectors.

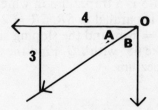

Example $\begin{pmatrix} -4 \\ -3 \end{pmatrix}$

Tan $A = \frac{3}{4} = 0{\cdot}75 \quad A = 36{\cdot}9°$ or $36°52'$
Acute angle with x-axis is $36{\cdot}9°$ or $36°52'$
Acute angle with y-axis is $53{\cdot}1°$ or $53°8'$

a) $\begin{pmatrix} -2 \\ -1 \end{pmatrix}$ x-axis b) $\begin{pmatrix} -4 \\ 3 \end{pmatrix}$ y-axis c) $\begin{pmatrix} -2 \\ -7 \end{pmatrix}$ y-axis

d) $\begin{pmatrix} 3 \\ -5 \end{pmatrix}$ y-axis e) $\begin{pmatrix} 2 \\ -6 \end{pmatrix}$ x-axis f) $\begin{pmatrix} -4 \\ 8 \end{pmatrix}$ x-axis

* 9 A ship sails 7 km on a bearing 081° and then 11 km on a bearing 048°. What is its bearing from the starting point? Give your answer to the nearest degree. (*Hint* Find its distance east (or west) and north (or south) of its starting point as in 13B question 21. From these the bearing can be calculated using tangent tables.)

* 10 An aeroplane flies 135 km on a bearing 142° and then 210 km on a bearing 013°. What is its bearing from its starting point? Give your answer to the nearest degree.

1st leg   **2nd leg**   **Total journey**

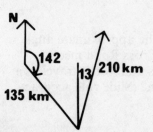
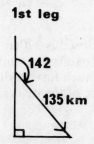

11 A balloon is launched and blown five miles by a north-east wind (i.e. a wind coming from the north-east). The wind then changes and it is blown 8 miles by a south-east wind. Find its bearing from its starting point (correct to the nearest degree).

12 A rectangle has its vertices at the points $(1, 2)$, $(1, 7)$, $(4, 2)$ and $(4, 7)$. What is the angle between a diagonal and the longer side?

13 Dick claims that he can run round the two sides of a rectangular playground while Molly is running across from corner to corner. The diagonal makes an angle of 30° with the longer side. If their speeds are in the ratio 6:5 (or 1·2:1), Dick being the faster, will he win his challenge?

14 What is the ratio of the height to the length of the base in an equilateral triangle?

15 Draw a geometrical figure that shows clearly that the tangent of an angle of 45° is 1. The figure should not be drawn to scale, but all the necessary dimensions should be marked clearly and a short explanation given.

16 Repeat question 15 to show that the sine of an angle of 30° (or the cosine of an angle of 60°) is $\frac{1}{2}$.

17 Using a sheet of 2 cm graph paper, draw the axes for the first quadrant only and mark them 0·1, 0·2, ... 1·0, using a scale of 20 cm to one unit. With the origin as centre, draw a circle of radius 20 cm. Draw in radius vectors at angles of 10°, 20°, 30° ... 80° with the x-axis, and call the points where they cut the circumference $P_1, P_2 \ldots P_8$. Read carefully the x and y co-ordinates of P_1, P_2 etc. and enter them in a table similar to the one below. In separate columns enter the values of sin and cosine of 10°, 20° etc., taking these values from 3-figure tables.

   If your drawing is sufficiently accurate, column 2 should agree with column 5, and column 3 with column 4.

1	2	3	4	5
Angle	x co-ordinate of P_1, P_2 etc.	y co-ordinate of P_1, P_2 etc.	Sine of angle	Cosine of angle
10°				
20°				
. . . . 80°				

Note The table illustrates the important principle that in a unit circle the x co-ordinate of the tip of the radius vector is the cosine of the angle swept out by the vector, and the *y* co-ordinate is the sine of this angle (the angles being measured anticlockwise from the positive direction of the *x*-axis).

18   A crane is used to lift a heavy box of treasure from the bottom of a dry well where it had laid concealed for centuries. The depth of the well is 20 m. If the jib of the crane is 8 metres long, and it is pivoted 1 metre above the ground, how much cable must be let out when the jib is inclined at
a) 50° to the horizontal,
b) 40° to the horizontal?
(Take the 'cable let out' as the length that hangs from the pulley downwards, and give your answer to the nearest tenth of a metre.)

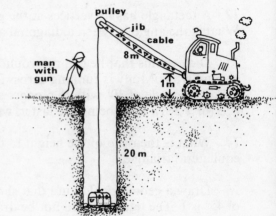

* 19   If no more than 25 metres of cable can be let out from the crane in question 18, what would be the maximum inclination of the jib of the crane? (To nearest degree.)

20   A class measured the heights of trees in the school playing field (which was almost level everywhere), using home-made clinometers. Tom held his clinometer at eye level (110 cm above the ground) and sighted it on the top of an oak tree, while Ali read the angle of inclination as 17°. Jane and Nancy held one end of the tape each, and found that Tom's distance from the oak tree was 45 metres. What was the height of the tree (to the nearest tenth of a metre)?

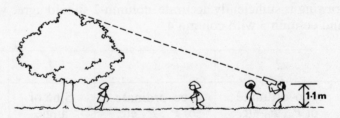

21   Jane and Susan, who were both about the same height as Tom, used their clinometer to observe an ash tree, and they found an angle of inclination of 22°. James and Robert measured the distance from where the girls were standing to the ash tree, and found it was 38 metres. What was the height of the ash tree?

22   They next turned their clinometer on a church steeple which, according to the vicar, was 40 m high, measured from the level of the churchyard. There was a drop of 3 metres going from the level of their playing field to the level of the churchyard. The angle of inclination was 33°. How far were they standing from the steeple? (Give your answer to the nearest metre.)

How to make a simple clinometer

Fairly long narrow tube

⊕ Thin cross wires stuck to one end

⊙ Other end covered with paper with pinhole in centre

Protractor stuck on with Bostik or Araldite. String with bob fixed to centre of protractor

Must be used by two people. One holds clinometer and takes a sighting. The other reads the angle.

The angle marked x is the angle to read

14 Pythagoras' Theorem

14A

Pythagoras' theorem states that 'in a right-angled triangle the square on the side opposite the right angle is equal to the sum of the squares on the other two sides'.

1 To verify the theorem for yourself, draw on graph paper a triangle *ABC* where angle $A = 90°$, $AB = 3$ cm and $AC = 4$ cm. Construct squares on all of the sides as shown in the diagram, and divide up the area of the largest square to simplify finding its actual area. The sum of the areas of the two smaller squares should be equal to the area of the largest one.

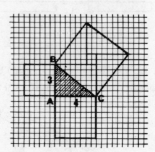

2 Use the method of question 1 to verify Pythagoras' theorem for a right-angled triangle of sides 4, 4. This time you will get four small triangles and no square.

3 Repeat question 1 with a triangle of sides 5, 4. This time you will get four triangles and a square 1×1.

4 Repeat question 1 with a triangle with sides 7, 5 (using a smaller scale). This time you will get four triangles and a square 2×2.

5 Repeat question 1 with a triangle 6×3. This time you will get 4 triangles and a square 3×3.

6 The property of a right-angled triangle which we have been studying above was known to the Chinese long before the days of Pythagoras. This is how they showed it:

O is the centre of the square, i.e. the intersection of its diagonals. The dotted lines are parallel and perpendicular respectively to the hypotenuse.

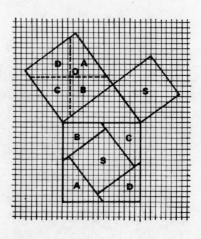

120

Inside the larger square you will see one of the smaller squares and four triangles. These triangles, fitted together, give the second of the smaller squares. The diagram is self-explanatory. On thin cardboard or thick cartridge paper draw a right-angled triangle as accurately as you can and complete the figure as shown. On another piece of paper copy the diagram exactly, and cut out the pieces marked A, B, C, D and S. They will fit together to give either the big square or the two smaller squares.

14B

1 On squared paper, using the first quadrant only and a conveniently large scale, join up the points O $(0, 0)$, A $(4, 0)$ and B $(0, 3)$ to form a right-angled triangle. Measure the length of the hypotenuse AB as accurately as you can, and verify by calculation that $OA^2 + OB^2 = AB^2$.

2 Repeat question 1 with the points $(0, 0)$ $(8, 0)$ $(0, 6)$

3 Repeat question 1 with the points $(0, 0)$ $(5, 0)$ $(0, 12)$

4 Repeat question 1 with the points $(0, 0)$ $(6, 0)$ $(0, 2·5)$

5 Repeat question 1 with the points $(0, 0)$ $(4, 0)$ $(0, 4)$

6 Draw the following right-angled triangles. In each case find the length of the hypotenuse by calculation and confirm by measurement:

   a) sides 4, 5   *b*) sides 3, 5   *c*) sides 4, 6   *d*) sides 5, 7   *e*) sides 8, 3.

Note When the lengths of all three sides of a right-angled triangle are whole numbers, these sets of three numbers are called 'Pythagoras' triples'. Examples are 3, 4, 5 and 5, 12, 13. These are studied in section 14F.

14C Square Roots

1 Use your tables to find the square roots of the following:

   a) 4·7   *b*) 47   *c*) 5·9   *d*) 59   *e*) 8·4   *f*) 3·6   *g*) 7·1   *h*) 29   *i*) 98   *j*) 6·6

2 Write down the square roots of the following without using tables:

a) 4	*b*) 400	*c*) 40 000	*d*) 36	*e*) 3600
f) 64	*g*) 6400	*h*) 640 000	*i*) 121	*j*) 12 100

3 Write down the square roots of the following without using tables:

a) 9	*b*) 10 000	*c*) 1600	*d*) 49	*e*) 250 000
f) 14 400	*g*) 900	*h*) 81	*i*) 640 000	*j*) 2500

4 Use tables where necessary to find the square roots of

   a) 1, 10, 100   *b*) 4, 40, 400

SEE! THAT'S WHAT COMES FROM HAVING SQUARE ROOTS!

5   In the questions above you can write down the square roots of numbers like 400 because $400 = 4 \times 100$ and $\sqrt{400} = \sqrt{4} \times \sqrt{100} = 2 \times 10$. Use this method to find the square roots of the following:

a) 500   b) 420   c) 350   d) 675   e) 828

6   Why is it not helpful to find the square root of 400 by using the fact that $400 = 40 \times 10$? What product would you use to find the square root of 4000?

7   Write each of these numbers as a product which will help you to find its square root:

a) 250   b) 3700   c) 1500   d) 7800   e) 780
f) 185   g) 18 500   h) 6500   i) 322   j) 3220

8   Using the tables and your products, find the square roots of each of the numbers in question 7.

9   Write down the square roots of the following using tables and writing suitable products where necessary:

a) 15   b) 15·5   c) 155   d) 4900   e) 490   f) 60   g) 600
h) 4800   i) 74·5   j) 7450   k) 720   l) 7200   m) 72 000   n) 1000
o) 102   p) 2·75   q) 27·5   r) 275   s) 2750   t) 13 200

10   Write down the square roots of the following:

a) 3·2   b) 32   c) 320   d) 3200   e) 32 000
f) 1·44   g) 14·4   h) 144   i) 1440   j) 520 000
k) 85   l) 810   m) 754   n) 75 400 o) 11
p) 110   q) 11 500 r) 1575   s) 32·96   t) 329·6

11   a) Look up the square root of 19. Suppose your answer was 4·359. This lies between 4 and 5. $4^2 = 16$ and $5^2 = 25$.   19 lies between 16 and 25, so your answer is probably correct. Use this method to check your answers to 9a), f), g), o) and s).

b) Suppose you found $\sqrt{740} = 27\cdot2$. This lies between 20 and 30. $20^2 = 400$, $30^2 = 900$. 740 lies between 400 and 900 so your answer is probably correct. Use this method to check your answers to 10c), d), i), l) and m).

c) Modify the above to check your answers to 10e), j), n) and q).
d) It is always good practice to check your square roots, but only use very simple checks that can be done in your head.

12   Repeat question 9 using logarithms.

13   Repeat question 10 using a slide rule.

14D

1   In the following examples you are
given the sides *a* and *b* of a right-
angled triangle. Find the hypotenuse *c*:

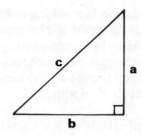

a) 14, 11   b) 7, 13   c) 21, 30   d) 40, 55   e) 9, 100
f) 2, 3   g) 40, 41   h) 225, 98   i) 35, 37   j) 11·2, 12·4
k) 6·95, 8·23   l) 21·4, 45·1   m) 4·55, 6·23   n) 7·77, 11·32

2   In the following examples you are given the hypotenuse *c* and one side *a*.
Find the other side *b*:

a) 12, 10   b) 23, 16   c) 8, 7   d) 134, 108   e) 67·3, 45·2
f) 224, 187   g) 77·3, 49·8   h) 665, 518   i) 38, 19   j) 2·35, 1·08

3   In the following examples you are given the three sides *a*, *b* and *c* in that
order. Find by calculation whether the triangles are right-angled. In each case
give your answer as 'Yes (exactly),' 'Yes (approximately)' or 'No':

a) 5, 12, 13   b) 9, 40, 41   c) 7, 8, 9   d) 7, 11, 12·65
e) 45, 60, 75   f) 4, 5, 6·40   g) 39, 80, 89   h) 21, 28, 35
i) 3, 6, 6·71   j) 4·2, 6·3, 8·4   k) 20, 40, 60   l) 7, 24, 25
m) 23, 30, 38

Confirm your answer to three of the questions by accurate drawing.

4   In this question all the triangles are isosceles right-angled triangles.
You are given the hypotenuse and you are required to find the lengths of the
two equal sides.

a) 100   b) 23·2   c) 55   d) 80   e) 11·85
f) 67·1   g) 7   h) 94   i) 39·5   j) 11

Confirm *a*), *b*) and *c*) by drawing (using a suitable scale).

14E Some of the Applications of Pythagoras' Theorem

These examples illustrate some of the kinds of problems that can be solved
by the application of Pythagoras' theorem. (In later books the theorem will
be used again in problems in trigonometry and in the geometry of the circle.)

1   A ladder standing on level ground against a wall just reaches a window
4·8 metres above the ground when its foot is pulled out 2 metres from the
wall. How long is the ladder?

2   A ladder is 5 metres long and stands on level ground. If it just reaches an
overflow pipe which is 4 metres above the ground, how far is its foot from
the wall?

3 A groundsman is marking out a tennis court on some newly laid tarmac. The length of the court is 23·8 metres and the width is 11 metres. He lays out a rectangle with these measurements, but is not quite certain that his angles are true right angles. So he checks by measuring the lengths of the two diagonals. They should be the same length. What should that length be (to the nearest 0·05 of a metre)?

4 A flagstaff stands on level ground outside an army headquarters. It is held by three wire guy-ropes. If these are fastened to the flagstaff 32 metres above the ground, and their lower ends are fixed to eyelets let into the concrete, these eyelets being 27 metres from the foot of the flagstaff, how much wire is used altogether in the three wire guy-ropes? (Answer to nearest metre.)

5 An aeroplane on a training flight flies 85 kilometres due east, and then turns and flies 105 kilometres due south. How far is it from its starting point?

6 Repeat question 5, this time the aeroplane having flown 48 kilometres south-west and then 93 kilometres north-west.

7 The diagram shows a right-angled triangle with sides 16 and 11 cm, the base being its hypotenuse. What is its area? What is the length of its hypotenuse? What is its height h?

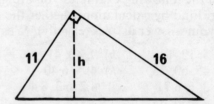

8 Repeat question 7 with sides 10 cm and 24 cm. Can you work the answer in your head without writing anything on paper?

9 The diagram shows two right-angled triangles and gives three dimensions. Calculate the length of side a.

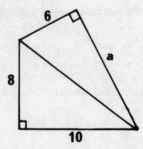

10 Repeat question 9 with lengths of 24 and 31 in the first triangle and 11 in the second.

11 The hypotenuse of an isosceles right-angled triangle is 5 cm. What are the lengths of the sides? What is its area?

12 Angela walks diagonally across a playing-field, 100 metres by 57·7 metres, while Tim runs round two sides, starting and finishing at the same points as Angela. How far did he run and how far did she walk?

13 A young man rowing a boat sets out to cross a river 30 metres wide, but the current carries him 15 metres downstream. How far does he row before reaching the far bank? (Answer to 0·1 metre.)

14 A boy is fishing in the canal on a calm day when there is no wind and virtually no current. His rod is extended to its full length of 6 metres and his line is hanging vertically at a distance of 4 metres from the bank. If the lower end of the rod is resting on the bank 60 cm above the level of the water and one metre from the edge, how far is the top of the rod above the water, assuming the rod is not bent? (Answer to 0·05 metre.)

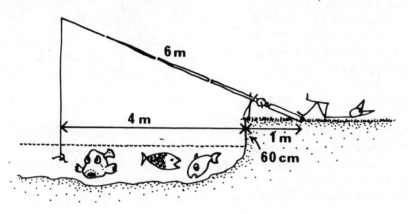

15 Find the distance between the following points whose co-ordinates are given:

Example (3, 1) and (7, 6)

Required distance (from the diagram) is

$$\sqrt{4^2 + 5^2} = \sqrt{41} = 6·40$$

a) (5, 4) and (8, 7)
b) (6, 3) and (12, 3)
c) (2, 5) and (−3, 8)
d) (7, 9) and (1, 4)   e) (4, −6) and (−3, 2)   f) (−1, −5) and (−3, −8)
g) (2, −6) and (−4, 3)   h) (−9, 11) and (2, 5)   i) (1, 3) and (4, 5)
j) (−6, 2) and (2, −6)

16 Find the lengths of the following vectors:

Example $\begin{pmatrix} 3 \\ 5 \end{pmatrix}$ Length = $\sqrt{3^2 + 5^2} = \sqrt{34} = 5·83$

a) $\begin{pmatrix} 4 \\ 7 \end{pmatrix}$ b) $\begin{pmatrix} 2 \\ 9 \end{pmatrix}$ c) $\begin{pmatrix} -3 \\ 1 \end{pmatrix}$ d) $\begin{pmatrix} 1 \\ 3 \end{pmatrix}$ e) $\begin{pmatrix} -3 \\ -3 \end{pmatrix}$

f) $\begin{pmatrix} 4 \\ -6 \end{pmatrix}$ g) $\begin{pmatrix} 5 \\ 2 \end{pmatrix}$ h) $\begin{pmatrix} 9 \\ -3 \end{pmatrix}$ i) $\begin{pmatrix} 12 \\ 12 \end{pmatrix}$ j) $\begin{pmatrix} 2 \\ -4 \end{pmatrix}$

17 These co-ordinates give the tips of position vectors. Find the lengths of the vectors:

a) $(3, 2)$ b) $(5, 4)$ c) $(11, -3)$ d) $(-6, -7)$ e) $(2, -1)$

18 A ship sails on a bearing 060° for a distance of 40 km, and then on a bearing 150° for a distance of 75 km. How far is it from its starting point?

19 A ship sails from A to B on a bearing 030° for a distance of 110 km, and then from B to C on a bearing 120° for a distance of 90 km. What is its distance from its starting point (to the nearest km)? What is its bearing from its starting point (to nearest degree)? (To answer the second question, calculate the angle CAB.)

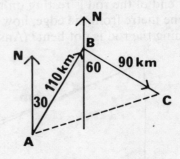

* 20 An aeroplane flies 340 km on a bearing 070° and then another 260 km on a bearing 220°. Find its bearing and distance from its starting point. *Hint* Calculate the distance travelled east (or west) and the distance north (or south) for each leg. Add these, and draw a right-angled triangle from which you can calculate the distance and bearing from the starting point.

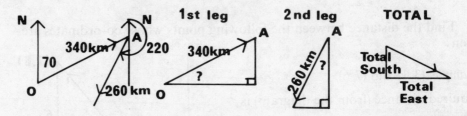

* 21 Repeat question 20, if the two legs of the journey are 130 km on a bearing 145°, and 300 km on a bearing 280°.

* 22 Repeat question 20 if the two legs are 90 km on a bearing 240° and 100 km on a bearing 030°.

* 23 The diagram shows a passageway with straight sides, a constant width of 2 metres, and a right-angled bend at BD. The problem is to find the longest ladder that can be carried round the bend without jamming, the ladder being horizontal.

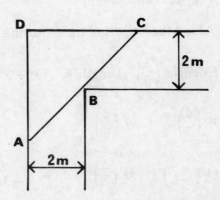

Our intuition tells us that jamming is most likely when the ladder is symmetrical to the corner, i.e. both AD and CD are of equal length 4 metres. But this must be investigated, not just assumed. Copy the sketch of the passageway on to graph paper with a suitable scale. Draw a series of lines ABC, with $AD = 10, 8, 6, 4, 3\frac{1}{2}, 3\frac{1}{4}$. Measure the length of AC in each case. You should find the least length is when $AD = 4$. This is the length of the longest ladder which can be carried round the bend.

Draw a graph to illustrate your results by plotting AD along the x-axis, and AC along the y-axis. You should get a curve, with the minimum value (lowest point) when $AD = 4$.

What is the length of the longest ladder that can be carried round the bend? (Answer to 0·05 metre.)

* 14F Pythagoras Triples

When the three sides of a right-angled triangle can all be expressed as integers, these integers are known as Pythagoras triples. We have already met the following: 3, 4, 5   5, 12, 13.

There are an infinite number of Pythagoras triples. Try and find all the triples in which the largest number is not greater than 100. There are 49 altogether.

1 As 3, 4, 5 is a Pythagoras triple, multiples of 3, 4, 5 are also Pythagoras triples, i.e. 6, 8, 10   9, 12, 15, etc. This will give you 20 triples to start with. List them in full.

2 5, 12, 13 is another well-known triple. This, with its multiples, will give you 7 more. What are they?

3 How many more triples can you find, not greater than 100? There are 22 more altogether.

Questions 4 onwards give you a systematic way of answering question 3. You may find them difficult, so they have been given two stars.

** *4* We have already seen that if we know the square of a number, the square of the next highest number is obtained by adding the sum of the two numbers to the square of the original number.
Thus $4^2 = 16$. To find 5^2 add $4+5$ to 16, and this gives 25.
But $4+5$ is 9 which is 3^2. So 3, 4 and 5 form a Pythagoras triple.

a) The next square after 9 is 16. But this can't be the sum of two consecutive numbers. Why not?
b) The next square is 25, which is the sum of 12 and 13. So 5, 12, 13 form a Pythagoras triple, as we already know.
c) 49 and 81 both give new triples. What are they?
d) You can get two more triples in this way. What are they?
e) Write down the 4 triples you found in c) and d) and also all their multiples not greater than 100. This will give you 8 more triples altogether to add to the 27 we found in questions 1 and 2, i.e. we now have 35 triples not greater than 100.

** 5 To find the square of a number two bigger than a given number, you add four times the middle number to the square of the smaller number. Thus

$$12^2 = 10^2 + (4 \times 11) \text{ i.e. } 144 = 100 + 44$$

If the middle number is a square, this gives a Pythagoras triple, the three numbers being 'middle number plus 1', 'middle number minus 1', and 'twice the square root of the middle number'.

Example Middle number 4: triples are 5, 3 and $(2 \times \sqrt{4}) = 4$

Try this method for middle numbers of 9, 16, 25 etc. You will get a lot of the triples we already know and three new triples.

a) What are these three new triples?
b) Write down the three new triples and their multiples not greater than 100. This gives eight new triples altogether, i.e. a total of 43 so far.

** 6 There are rules similar to those in questions 4 and 5 for the squares of numbers differing by 3, 4, 5, 6, 7, but none of these give new triples, only triples we already know.

There are, however, three new triples for numbers differing by 8. The rule here is 'square of larger number equals square of smaller number plus (16 × middle number)'.

Example $20^2 = 12^2 + (16 \times 16)$ or $400 = 144 + 256$

So if the middle number is a perfect square, the three numbers will form a Pythagoras triple, the three members of the triple being: (middle number + 4), (middle number − 4) and (4 × square root of middle number).

Taking the middle number as 9, this gives 13, 5 and (4 × 3), i.e. 13, 5, 12, a triple we already know. Using this method, however, gives three new triples.

a) What are these three triples?
b) Write down the triples and their multiples not greater than 100. This adds another five to our triples, giving 48 so far.

** 7 The next triple comes from two numbers differing by 9.
The rule here is 'square of larger = square of smaller
plus (9 × sum of numbers)'.
Thus $19^2 = 10^2 + (9 \times 29)$ or $361 = 100 + 261$.
So if you can find a pair of numbers differing by nine, such that their sum is a perfect square, you can form a Pythagoras triple as follows: the two numbers and (3 × square root of sum of numbers).

Example 8 and 17. The sum of the numbers is 25, which is the square of 5. So the triple will be 8, 17 and (3 × 5), i.e. 8, 17, 15. But this triple we already know. This method however throws up two new triples, taking our total to 50. What are the new triples?

8 This method now becomes extremely tedious but two new triples can be found, one for numbers differing by 18 and one for numbers differing by 25. What are they? These bring our total of triples, with no number exceeding 100, to 52.

9 Make a list of the Pythagoras triples you have found. There should be 52 of them, but list the ones you have, even if there are less than the full 52.

Arrange each triple with the smallest number first and the largest number last, the smallest numbers getting steadily bigger.

Start of table:

3	4	5	
5	12	13	
6	8	10	
7	24	25	
8	15	17	
9	12	15	etc.

There are other ways of calculating Pythagoras' triples. Two of these will be given in a later book.

There is an infinity of right-angled triangles whose sides form Pythagoras triples. Yet the right-angled triangles whose sides do not form Pythagoras triples are far more numerous!

CONGRATULATIONS, MR. PYTHAGORAS!

15 Infinite Decimals, Recurring and Non-recurring

15A Recurring Decimals

1 Express the following fractions as decimals (all of which terminate):

a) $\frac{1}{2}$ b) $\frac{1}{4}$ c) $\frac{1}{5}$ d) $\frac{1}{8}$ e) $\frac{1}{10}$ f) $\frac{1}{16}$

g) $\frac{1}{20}$ h) $\frac{1}{25}$ i) $\frac{1}{32}$ j) $\frac{1}{40}$ k) $\frac{1}{50}$ l) $\frac{1}{64}$

(*Hint* To get $\frac{1}{16}$ divide $\frac{1}{8}$ (0·125) by 2. Similarly for the other harder fractions.)

2 The denominators of all the fractions in question 1 are multiples of one or both of *two* primes, and of no other number. What are these two primes?

3 Express the following fractions as decimals, all of which recur. Some recur on a single figure, e.g. 0·4444 . . . which can be written 0·4̇. Some recur on a group of figures, e.g. 0·234523452345 . . . which can be written 0·2̇345̇. All of the decimals recur on groups of 6 figures or less.

a) $\frac{1}{3}$ b) $\frac{1}{7}$ c) $\frac{1}{11}$ d) $\frac{1}{13}$ e) $\frac{1}{37}$ f) $\frac{1}{41}$

* 4 Repeat question 3 for the following fractions, all of which give decimals that recur. (The figure in brackets after the fraction gives the number of figures in the recurring group.)

a) $\frac{1}{17}$ (16)   b) $\frac{1}{19}$ (18)   c) $\frac{1}{23}$ (22)   d) $\frac{1}{29}$ (28)   e) $\frac{1}{31}$ (15)

5 What do you notice about the numbers in the denominators of the fractions in questions 3 and 4?

6 Change the following fractions to decimals. None of the recurring decimals have groups of more than 8 figures, but use a calculating machine if possible.

a) $\frac{2}{3}$ b) $\frac{4}{25}$ c) $\frac{3}{7}$ d) $\frac{3}{11}$ e) $\frac{7}{128}$ f) $\frac{5}{13}$ g) $\frac{6}{37}$

h) $\frac{1}{200}$ i) $\frac{8}{75}$ j) $\frac{7}{625}$ k) $\frac{16}{101}$ l) $\frac{22}{137}$ m) $\frac{28}{239}$

7 Work out as decimals: $\frac{1}{7}$ $\frac{2}{7}$ $\frac{3}{7}$ $\frac{4}{7}$ $\frac{5}{7}$ $\frac{6}{7}$
Each of these gives a decimal that recurs on a group of six figures.
What do you notice about the six groups of six figures?

8 Repeat question 7 for the thirteenths, i.e. $\frac{1}{13}$ $\frac{2}{13}$ . . . $\frac{12}{13}$
Once again each fraction gives a recurring group of six figures.
What do you notice about the twelve groups of six figures?

9 Calculate the 41st's, i.e. $\frac{1}{41}, \frac{2}{41} \ldots \frac{40}{41}$
If you have a calculator, find all 40. They recur on groups of 5 and there are
eight different groups. If you do not have a calculator, try and find
the eight groups by long division. You will not have to go beyond $\frac{15}{41}$.

10 Calling the 41st's 1, 2, 3 etc. and the groups A, B, C, show which 41st
uses which group. Thus $1A$, $2B$, ... $8E$ etc. (If you have calculated only a
few groups, do it for the ones you have calculated).

✳ **11** Check as many as you can of the following:

    a) The 37ths cycle on 12 groups of 3 figures.
    b) The 101sts cycle on 25 groups of 4 figures.
    c) The 73rds cycle on 9 groups of 8 figures.
    d) The 53rds cycle on 4 groups of 13 figures.
    e) The 31sts cycle on 2 groups of 15 figures.

In each case try and find two or three of the groups, but do not attempt to
find them all.

✳ **12** The preceding examples show that if P is a prime (other than 2 or 5)
then $\frac{1}{P}$ $\frac{2}{P}$ $\frac{3}{P}$ expressed as decimals form recurring groups which either contain
$(P-1)$ figures or a factor of $(P-1)$. The product 'number of groups' $\times$
'number of figures in a group' always gives $(P-1)$. Check this for the examples
in questions 7, 8, 9 and 11.
    (In general this rule does not work if P is not prime.)

Fractions with denominators which are composite, i.e. have factors

✳ **13** Change the following fractions to decimals, all of which recur on groups
of eight figures or less.

    a) $\frac{1}{9}$ *b)* $\frac{1}{21}$ *c)* $\frac{1}{27}$ *d)* $\frac{1}{33}$ *e)* $\frac{1}{39}$

(*Hint* These decimals are easily calculated from the decimals in question 3,
e.g. $\frac{1}{21}$ is obtained by dividing $\frac{1}{7}$, i.e. $0 \cdot 14285714285714 \ldots$ by 3.

✳ **14** The following fractions give decimals that are partly recurring, i.e. the
first one, or two, or three ... figures do not recur, and the remainder recur
singly or in groups, e.g. $0 \cdot 143587958795879 \ldots$ which can be written $0 \cdot 143\dot{5}87\dot{9}$.
    The first figure in brackets after each fraction gives the number of non-
recurring figures, and the second the number of figures in the recurring group.
Change each fraction to a decimal, using the hint in question 13.

    a) $\frac{1}{12}$ $(2,1)$ *b)* $\frac{1}{14}$ $(1,6)$ *c)* $\frac{1}{15}$ $(1,1)$ *d)* $\frac{1}{22}$ $(1,2)$ *e)* $\frac{1}{24}$ $(3,1)$

✳ **15** What difference do you notice between the numbers in the denominators
of the fractions in question 14 and those in question 15?

16 We can now summarise the results we have found when turning fractions into decimals:

		Factors of denominator	Type of decimal	Example
	a)	2's and/or 5's only	Terminating	$\frac{1}{160} = 0.00625$
	b)	No factors: prime	Recurring	$\frac{1}{13} = 0.\dot{0}7692\dot{3}$
*	c)	Factors excluding 2 and 5	Recurring	$\frac{1}{63} = 0.\dot{0}1587\dot{3}$
*	d)	Factors including 2 and/or 5 and one or more other primes	Partially recurring	$\frac{1}{48} = 0.0208\dot{3}$

Give two examples of each type, using 2, 3 or 4 in the numerator and fairly small numbers in the denominator.

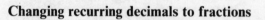
MATHEMATICAL FACTS, NUMERICAL REACTIONS CHANGE THESE DECIMALS INTO FRACTIONS!

Changing recurring decimals to fractions

17 Write down the factors of   a) 111   b) 1111   c) 11 111   d) 111 111

18 Change the following recurring decimals to fractions and cancel down to the lowest possible terms.

a) $0.\dot{3}$ b) $0.\dot{5}$ c) $0.\dot{6}$ d) $0.\dot{1}\dot{8}$ e) $0.\dot{6}\dot{3}$ f) $0.\dot{8}\dot{1}$ g) $0.\dot{1}0\dot{8}$

Example $0.\dot{4} = \frac{4}{9}$

* 19 Change the following recurring decimals to fractions and cancel down to the lowest possible terms, using the factors you found in question 18.

a) $0.\dot{2}\dot{7}$ b) $0.\dot{3}\dot{6}$ c) $0.\dot{8}\dot{4}$ d) $0.\dot{1}3\dot{8}$ e) $0.\dot{4}7\dot{1}$

f) $0.\dot{8}5\dot{5}$ g) $0.\dot{1}78\dot{2}$ h) $0.\dot{8}3763\dot{}$ i) $0.\dot{0}31746\dot{}$ j) $0.\dot{0}54945\dot{}$

Example $0.\dot{0}2\dot{7} = \frac{027}{999} = \frac{27}{999} = \frac{3}{111} = \frac{1}{37}$

* 20 Change the following partially recurring decimals to fractions.

a) $0.1\dot{3}$ b) $0.1\dot{6}$ c) $0.2\dot{4}$ d) $0.5\dot{7}$ e) $0.02\dot{3}$

f) $0.12\dot{6}$ g) $0.01\dot{8}$ h) $0.11\dot{2}$ i) $0.11\dot{2}$ j) $0.10\dot{5}$

Example $0.2\dot{4}\dot{5} = N$ $10N = 2.\dot{4}\dot{5} = 2\frac{45}{99} = 2\frac{5}{11} = \frac{27}{11} \therefore N = \frac{27}{110}$

* 21 Change the following partially recurring decimals to fractions.

a) $0.8\dot{6}$ b) $0.08\dot{3}$ c) $0.83\dot{3}$ d) $0.13\dot{6}$ e) $0.31\dot{8}$

f) $0.5\dot{2}2\dot{5}$ g) $0.50\dot{5}9\dot{4}$ h) $0.1\dot{4}2857\dot{1}$ i) $0.10\dot{3}1746\dot{}$ j) $0.1\dot{7}1428\dot{5}$

15B Infinite Decimals, Non-recurring: Enlarging the Number Field

1 So far we have encountered two sets of numbers: *a*) the integers, with positive and negative integers as subsets (the positive integers also being known as natural numbers); *b*) the rationals, with positive and negative rationals as subsets. Give three examples of each of *a*) and *b*).

2 We now meet a third type of number which is neither an integer nor a rational, and is called *irrational*.

We have already seen that every rational number can be represented by a decimal, and that this decimal either terminates or (sooner or later) recurs. We have also seen that every decimal that either terminates or recurs can be represented as a fraction, i.e. as a rational number. Can you make up a decimal which does not terminate (i.e. it goes on and on to infinity) and does not recur? Give more than one example if you can. It will be an irrational.

3 Now consider $\sqrt{2}$. The value of $\sqrt{2}$ is given in the tables as 1·414. If you square 1·414 you get 1·999396. This is nearly equal to 2, but not exactly. It is a little too small. Obviously 1·414 is not the exact square root of 2.

   a) So try squaring 1·4141, 1·4142, 1·4143, etc. You will soon come to a number whose square is greater than 2. The square root of 2 then lies between this number and the one before it. This gives you the figure in the fourth decimal place of $\sqrt{2}$.
   b) Now try in the same way to get the figure in the fifth decimal place. Use a calculating machine if possible, but you will soon run out of capacity.
   c) Get as many more places as you can.

4 The tables give $\sqrt{3}$ as 1·732. Use the method of question 3 to find as many more figures as you can.

5 Repeat question 4 with $\sqrt{5}$ (given in the tables as 2·236.)

6 There are many better methods of finding square roots. (One such method is given in 16B). Using electronic computers, the square root of 2 has been worked out to hundreds of decimal places, but it does not terminate and it does not recur. In the light of question 2 above, what can you deduce about $\sqrt{2}$? Does the same hold for $\sqrt{3}$ and $\sqrt{5}$?

* *7* If you have given the correct answer to question 6, you will now know that $\sqrt{2}$, $\sqrt{3}$ and $\sqrt{5}$ are irrationals, i.e. they cannot be represented in the form $\frac{p}{q}$ where *p* and *q* are integers. Referring to the Farey lattice the line representing $\frac{3}{4}$ goes through the lattice point (4, 3). It also goes through (8, 6).

   a) Name three other points that this line goes through.
   b) How many lattice points does it go through altogether?
   c) How many lattice points does the line representing $\sqrt{2}$ go through?
   d) How many lattice points do the lines representing $\sqrt{3}$ and $\sqrt{5}$ go through?
   e) How many lattice points does the line representing any irrational go through?

The fact that $\sqrt{2}$ is not a rational, i.e. cannot be expressed in the form of $\frac{p}{q}$ where p and q are integers, was well known to the Greeks, who produced a simple proof, which is given in 16B.

8 Irrationals occur in solving algebraic equations.
Thus if $x^2 = 2$, $x = \pm\sqrt{2}$.
Which of the following equations have rational roots and which have irrational roots?

a) $9x^2 = 4$ b) $9x^2 = 7$ c) $8x^2 = 32$ d) $x^2 - 5 = 0$

e) $x^2 - 4 = 0$ f) $16x^2 = 32$ g) $9x^2 - 27 = 0$ h) $9x^2 - 27 = 9$

i) $4x^2 - 20 = 0$ j) $3x - 7 = 2$ k) $5x = 2\sqrt{2}$ l) $3x = 6$

16 Everyday Arithmetic and Exploring More Byways

16A Everyday Arithmetic

1 Robert rides a bicycle to school five times a week, forty weeks in the year. His home is about 12 km from his school. He averages about 80 km a weekend and during the 12 weeks' holiday in the year he averages 140 km a week (including weekends).

What is the approximate distance he travels in a full year, and what is the average distance in a week?

2 Robert insures his bicycle for £1 a year. This covers accidental damage and in the event of loss or total destruction, the original cost less depreciation. Depreciation is calculated as 15% of the original purchase price of the bicycle in the first year, and 10% of the original purchase price in each succeeding year. If Robert's parents originally paid £48 for the bicycle, calculate:

a) its value after one year   *b)* its value after two years

c) its value after five years.

3 Robert uses his bicycle in all weathers on very rough roads so the wear and tear is unusually heavy. The following table gives the number of replacements he has to purchase over a five year period and the approximate cost of each replacement:

Item	No. of replacements over five years	Cost per replacement
Tyres	4	£3.00
Inner tubes	4	£1.10
Chain	2	£2.00
Brake blocks and shoes	6 pairs	32p a pair
Saddle cover	1	£1.00
Saddle	1	£3.25
Saddle bag	1	£3.50
Bell	1	65p
Pair of mudguards	1	£2.85 a pair
Rear reflector	1	36p
Bulbs for electric (dynamo) lamps:		
Front	2 ⎫	
Rear	2 ⎭	15p each
Lubricating oil (tins)	8	32p a tin
3-speed cable	1	£1.10

a) Find the total cost of these items over a five year period.

b) Adding the depreciation from question 2, and the cost of insurance, what was the total cost of running the bicycle for five years?
c) What was the annual cost?
d) Using the results of question 1, what was the cost per km?
e) What was the cost per week *i)* excluding depreciation, *ii)* including depreciation?

4 Robert's sister Emily averages only 20 km a week, and her bill for replacements is nothing. However, she has an accident in which the front wheel is severely buckled. A new wheel costs £4.95, but she only has to pay the first £1 of this as the insurance company pay the rest. Over a five year period she buys two tins of lubricating oil. Her bicycle originally cost £52.

a) What was the depreciation over a five year period?
b) What were her total other costs?
c) What was the approximate distance she covered each year?
d) What did her bicycle cost per km to run?

5 A young man buys a power-assisted bicycle which is marked 'Cash Price: £130'. He pays £20 deposit and 12 monthly instalments of £11.

a) How much does he pay altogether?
b) By how much does this exceed the cash price, i.e. how much interest does he pay?
c) How much did he owe after he had paid the deposit?
d) Express the interest as a percentage of the sum in *c*).

6 *a)* Mrs McDonald bought two dozen large eggs at 50p a dozen, 3 packets of butter at 25p a packet, 3 kilograms of sugar at 22p a kilogram, 2 tins of condensed milk at 19p a tin, 2 pots of jam at 30p a pot, and 2 bags of self-raising flour at 26p a bag. What was her total bill?
b) If a year later the above prices had risen to 57p, 27p, 24p, 20p, 32p, and 28p respectively, what would her total bill have been for the same purchase?
c) If after another year the prices were 57p, 28p, 25p, 21p, 33p and 29p, what would her total bill have been this time for the same purchase?
d) Express the increase in her bill over the first year as a percentage of the total price she paid at the beginning of that year.
e) Express the increase in her bill during the second year as a percentage of the total price she paid at the beginning of that year.
f) Do these two percentages represent the rate of inflation in the two years? Discuss.

7 A lock on the Grand Union Junction canal holds water to a depth of 3 metres when full. Its width is $4\frac{1}{2}$ metres and its length 24 metres.

a) What volume of water does it hold when full? Give your answer in
   i) cubic metres   *ii)* litres.
b) When the paddles on the lower gate are opened, the lock starts to empty. When the water level has dropped by $1\frac{1}{2}$ metres, how much water has flowed out?

c) If it takes three minutes for the water level to drop by $1\frac{1}{2}$ metres, at what average rate is the water running out? Give your answer
    i) in cubic metres per minute   *ii*) in litres per second.

8 A mother bakes her own bread. Using 450 g of strong white flour and 450 g of strong brown flour, two teaspoonsful of salt and a packet of dried yeast, she gets two loaves of equal weight. Strong flour (white or brown) costs 34p for a bag holding 1·5 kg, dried yeast is 6p a packet, and the cost of the salt is trivial. Cooking costs nothing, as the cooker is always hot whether she is baking bread or not.

    a) What is the cost of one loaf?
    b) If a loaf of sliced bread in the baker's costs 22p and weighs about the same as a home-made loaf, what does she save on each loaf?
    c) What percentage saving is this, reckoned as a percentage of the shop price?

9 Marian and Danny, her brother, invite ten friends each to a joint birthday party. The whole party is not to cost more than £10.
    They decide to buy 2 dozen balloons at 15p a dozen; 10 prizes at an average of 5p each, ten at an average of 10p each, and one at 50p; 20 'go home' presents at 10p each; 2 boxes of chocolates at 65p each for two friends who will help at the party; 1 kg of sausages (60p for $\frac{1}{2}$ kg); 2 tins of pineapple at 35p (special offer) and $\frac{1}{2}$ kg cheese (60p); 1 packet of cocktail sticks (18p); 4 litres of ice cream costing £1.50 and 50 cornet cases at 25p; 2 tins of sardines at 22p a tin; two sliced loaves at 22p each; and 1 packet of candles at 18p.

    a) What was the total cost of the things the children had listed?
    b) By how much per cent had they exceeded the £10 they were allowed to spend?

10 Judith gets up at 7.15 am and sets off for school at 8.15. She reaches school at 9 o'clock. School ends at 3.45 pm. She arrives home at 4.30 pm. After having her tea she watches the children's programmes on television from 5 pm till 6 pm, and then takes the dog for a walk if it is still light. At 6.30 she starts her homework and works till 8 pm. Then she watches television till 9 pm, and talks to her family or reads until 10 pm and goes to bed. She is allowed to read till 11 pm, and then it is 'lights out' and she falls asleep at once. Answer the following questions:

    a) How many hours a day does she spend in bed?
    b) How many hours a day does she spend travelling to and from school?
    c) How many hours a day is she at school?
    d) How many hours a day does she spend on homework?
    e) How many hours a day does she spend watching television?
    f) How much time is left?
    g) Express your answers to questions *a*) to *f*) as percentages of a whole day (to the nearest $\frac{1}{4}$%).

11 Charles and his father decide to lay a concrete path down one side of their garden. They dig out a shallow trench about 80 cm wide and 20 cm deep, shutter the sides of the trench with boards salvaged from an old shed, and fill the trench with broken brick (of which they have plenty) to a mean depth 4 cm below the top of the shuttering. Up to this point the job has taken them about twelve hours.

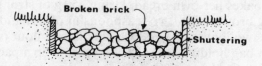

a) The length of the path is 22 metres, the width between the inner faces of the shuttering is 80 cm and the depth of the concrete is to be 4 cm. How much ready-mixed concrete do they need?
Here they run into a difficulty, as their suppliers do not normally deliver less than one cubic metre, but they eventually agree to deliver $\frac{3}{4}$ of a cubic metre provided the full price for a cubic metre is paid, i.e. £28.
b) It takes them one hour to lay the concrete. If the builder had quoted £55 for the whole job, would this have been a reasonable price?
c) How much did they save by doing the job themselves?

12 A young couple buy an old house, which is in need of extensive repair. One of the most urgent jobs is to fit a new floor to an upstairs bedroom which is 5 metres square. The joists are sound but the floorboards are rotten. They decide to use standard 'tongue and groove' floorboards, whose effective width is $11\frac{1}{2}$ cm.

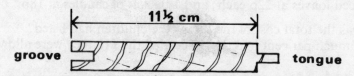

a) How many metres should they order, if they allow 5% extra for wastage in cutting up? (Answer to the nearest 5 m above.)
b) If they need 7 kg of nails at £1.20 a kilogram, and if the floor board costs 40p a metre, what will be the total cost of the floor?
c) If they can lay four boards a day in the spare time available, how long will it take them to complete the floor?
d) If they decide to replace the skirting board, and this costs 60p a metre, how much will this add to the cost of the job, including an extra half kilogram of nails?

13 The young couple in question 12 decide they cannot afford to buy a carpet just yet, so they leave their new floor for three months to season and then give it two coats of light oak stain at a cost of 95p a coat. They then give it 2 coats of polyurethane clear varnish, this costing £1.05 a tin, and one tin holding enough varnish to cover 18 square metres of floor. They use half a tin each of primer, undercoat and gloss to paint the skirting board white. The tins cost respectively 95p, £1.05 and £1.10, but the unused half tins can be used for another job. What is the total cost of staining and varnishing the floor and painting the skirting board?

14 Mr Martin decides to make
a new gate for his drive.
Here is his design.

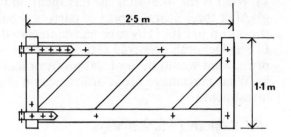

He plans to use 13 cm by 6·5 cm planed timber at a cost of £2.25 a metre.

a) Taking the length of the sloping pieces to be 1·5 times the length of an upright side, what is the total length of timber he requires?
b) What will be the cost of the timber, allowing an extra 5 % for wastage in cutting and ordering to the nearest metre above?
c) A hinge set costs £5, a self-closing fastener costs £3, heavy chain link to protect the front of the gate costs £2, two new oak gate posts cost £9.80 each, aluminium primer for these two gate posts, and paint and primer for the gate and posts costs altogether £3. All the 10 joints are morticed and are then bolted as well as glued. He uses one container of glue costing 95p, and 10 bolts costing 15p each. He also needs 4 bolts for the top hinge and 2 for the bottom, again costing 15p each.
What is the total cost of the gate?

15 Deborah buys George, a parrot, for £30, plus £10 for his cage.
She pays for George and his cage by selling some savings certificates she had been given. Her father builds a large outdoor cage for George at a total cost of £25, which Deborah also pays.
Deborah feeds George on oranges, parrot seed and artificial nectar. George consumes an orange a day, which costs, on average, 5p per orange. A bag of parrot seed costs £1 and lasts 6 months. Once a week the nectar is blended and stored in the refrigerator. Here are the ingredients:

Ingredient	Cost per tin, jar or packet	Number of weeks it lasts
Baby rusks	35p	4
Honey	42p	5
Peanut butter (smooth)	45p	6
Sweetened condensed milk	32p	5
Baby cereal	65p	52
Baby rice	24p	26
Glucose	30p	30

a) Calculate the cost of the nectar per week (to the nearest tenth of 1p).
b) Calculate the cost of the oranges per week.
c) Calculate the cost of the parrot seed per week.
d) What is the total cost per week of feeding George?
e) What is the cost per year, if Deborah spends £1 a year on extras such as millet sprays?

f) What is the total sum she has spent on George in three years?

g) After three years, import of parrots is banned, and George's value shoots up to £100. His cage has depreciated to £5 and the outdoor cage to £12. If Deborah now sold George and his two cages for the proper market price, what would be her total profit or loss over the three years?

h) What percentage profit or loss would she have made on her total investment?

16B Exploring More Byways

1 *a*) Multiply 111 111 111 by 111 111 111. The result will intrigue you.

   b) Find the square root of 123 454 321.

2 'I would like an extra week's holiday this year, madam,' said the gardener. 'How many hours a day do you work, Dodson?' asked the Duchess, smiling. 'Eight, madam.'

'That's a third of a day. There's 366 days this year. One third is 122 days. But then you have Saturday and Sunday off every week. That's 104 days in the year. That leaves 18 days that you work. Then you already have two weeks' holiday, that's fourteen days,' the duchess went on, 'and you have four bank holidays a year, so in actual practice you don't work at all.'

Dodson took out his calculator. 'I think you've got your figures wrong, madam. I have Saturdays and Sundays off every week, fourteen days' holiday and four bank holidays a year. That's 122 days off, just one third of the year. So I must work two thirds of the year and that's sixteen hours a day. But the regulations only allow me to work eight hours a day without being paid overtime, so it looks as though you owe me a lot of overtime, madam.'

Dodson had his extra week's holiday. But where did their arguments go wrong? Discuss.

3 The Ancient Greeks used a simple form of logarithms. This is how they worked:

1	2	3	4	5	6	7	8	9	10
2	4	8	16	32	64	128	256	512	1024

11	12	13	14	15	16
2048	4096	8192	16 384	32 768	65 536

To multiply any two numbers in the bottom row, they added the numbers in the top row, and looked at the corresponding number in the bottom row.

Example To multiply 32 by 16. Add 5 and 4 to give 9. The number under 9 is 512. This is the answer.

   a) Continue the table as far as 20 in the top row. What are new numbers in the bottom row?

   b) Use it to find the following products: *i*) 32×64 *ii*) 32×256

   iii) 128^2 *iv*) $\sqrt{65536}$ *v*) $4096 \div 64$ *vi*) 512×2048

   c) The logarithms we used in section 5 were logarithms to base 10. In base 10, because $10^2 = 100$, log 100 is 2. To what base were the logarithms calculated in the table above?

* 4 Many of the world's finest mathematicians have made great efforts to solve some of the problems concerning prime numbers. In particular they have tried to find

1 a simple test for saying whether or not a large number is prime;
2 a simple way of writing down the number of primes between two given numbers;
3 simple ways of 'constructing' primes.

One famous conjecture was that $n^2 - n + 41$ was always prime.

a) Why is this obviously wrong for $n = 41$?

b) Work out the numbers obtained when $n = 5, 7$ and 11 and check whether or not they are prime.

c) If you have calculating equipment, repeat question *b)* for $n = 1$ to $n = 40$, arranging your numbers in sets of 5 for easy reference and checking your answers with a table of primes (if you have access to such a table).

d) If $n = 42$, is the number obtained a prime?

e) Repeat *d)* for $n = 43$.

* 5 Another famous conjecture about primes was that numbers of the form $2^p - 1$, where p is prime, were prime numbers for $p = 2, 3, 5, 7, 13, 17, 19, 31, 67, 157$ and 237. These numbers are known as *Mersenne numbers* after Father Mersenne who made the conjecture.

a) Work out the Mersenne numbers for $p = 2, 3, 5, 7, 13$.

b) Are they prime?

c) Work out the Mersenne number for $p = 11$. Eleven was not included in Mersenne's list so it should have factors. Find them.

** *d)* M_{67} proved not to be prime. Its factors were found by F. N. Cole in 1903, after working for 'three years of Sundays'.

The factors are 193,707,721 and 761,838,257,287.

If you like to multiply these numbers longhand, and also work out the value of $2^{67} - 1$, the two answers should agree. An ordinary desk calculator won't help very much, although the calculation would be easy on a really large computer.

6 Two other conjectures about primes, both since proved, are as follows:

a) Any prime of the form $4n + 1$ can be expressed as the sum of two squares, and this can be done in one way only.

Thus the first primes of this form are 5, 13, 17.

5 is $2^2 + 1^2$ and 13 is $3^2 + 2^2$.

i) What is 17?

ii) Work out the next four primes of the form $4n + 1$ and find the two squares for each.

b) Any square number except 1 can be written as the sum of two primes, but this is not necessarily unique, i.e. it can often be done in more than one way.

Example $2^2 = 4 = 1 + 3$ $\quad 3^3 = 9 = 7 + 2$

i) Check this assertion for the next five squares.

ii) Give three ways of expressing 10^2 as the sum of two primes.

* 7 There exist whole blocks of numbers that contain no primes. Thus there are no primes between 241 and 251. One of the most interesting types of 'no prime' blocks is $n! + 2$, $n! + 3 \ldots n! + n$.

$n!$ is called factorial n and means $1.2.3\ldots n$. Thus factorial 5 is $1.2.3.4.5$. i.e. 120. So the block of numbers 122 to 125 contains no primes.

a) Check this for yourself. Check also the block for factorial 6, i.e. 722 to 726 and for factorial 7, i.e. 5042 to 5047. This is not as difficult as it looks because one of the integers smaller than n is a factor each time.

b) Can you see why $n! + p$ where p is an integer and $2 \leqslant p \leqslant n$ is always composite, i.e. always has factors?

c) $n! + 1$ is not included in the list.

Is it prime or composite?

Test it for $n = 2, 3, 4, 5, 6, 7$.

8 You are given twelve apparently identical billiard balls, but one is heavier or lighter than the other eleven. You are given a pair of scales with two pans, but no weights. The problem is to find the odd ball and whether it is heavier or lighter than the rest, and to do this in the smallest possible number of weighings.

a) Can you do it in 11 weighings? Give the details.

b) Can you do it in 6 weighings? Give the details.

** *c)* Can you do it in 5 weighings? Give the details.

d) Can you do it in 4 weighings? Give the details.

** *e)* Can you do it in 3 weighings? This is the smallest possible number of weighings.

9 How to cheat with statistics

The use of statistics, honestly applied, helps to make facts clear and easily understood. But if used unscrupulously, statistics can be very misleading.

Here are two examples.

a) A firm increased its sales of a product in three successive years from £100 000 to £110 000 and £120 000. Honestly shown on a bar chart this would be represented as Fig. 1 below. But if a 'false zero' is used, it would look as though the sales had doubled and trebled in the second and third year, as in Fig. 2.

Fig. 1

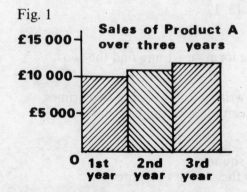

Fig. 2

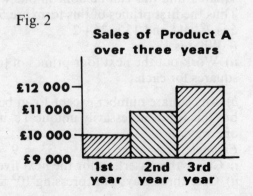

142

If Fig. 2 had been labelled 'Excess sales over £90000 a year', then it would not have been so misleading.

b) Three firms made different sizes of tinned goods, all the tins being the same shape but of different heights. If the heights were in the ratio $1:1\cdot1:1\cdot2$, and the firm making the smaller tins used this ratio to draw a bar chart comparing their products with the products of the other two firms (Fig. 3), it would give a misleading picture; whereas a true picture would be to compare the volumes, which are in the ratio $1^3:1\cdot1^3:1\cdot2^3$, i.e. $1:1\cdot33:1\cdot73$ as shown in Fig. 4.

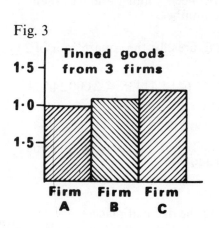

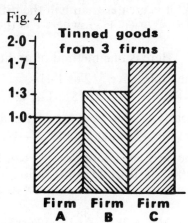

c) Think up an example similar to a) comparing the examination results of four pupils.

d) Think up an example similar to b) comparing the sizes of four Easter eggs which are hollow and filled with wrapped chocolates.

e) Can you think of an example of another type of misleading statistics?

10 The numbers in the diagram form a 'magic square'. The sum of the numbers in each row, each column and each diagonal is the same (15). Only the numbers 1 to 9 have been used.

4	3	8
9	5	1
2	7	6

a) If we ignore the restriction that only the numbers 1 to 9 can be used and use any positive integers, it is easy to make up magic squares indefinitely. Here is the start of such a square. Can you complete it:

2		
5		10

b) Here are four more squares for you to complete:

4		
5		10

6		
3		8

12		
7		14

7		
6		11

c) Comparing your completed squares
in *a*) and *b*) with the square opposite,
what do you notice about *i*) x and y
ii) the middle number in relation to x and y
iii) the sums of the rows in relation to
the middle number?

d) Make up some squares for yourself, choosing x and y first and using the
results you found in *c*) above. To avoid negatives, choose the value of z so
that it is greater than half the difference between x and y and less than
$(x + e)$ or $(y + e)$, where e is the middle number.

** **11** Here is one method of finding extra figures for the square root of 2

The tables give $\sqrt{2}$ as 1·414.

$1·4141^2 = 1·999\,678\,81$; $1·4142^2 = 1·999\,961\,64$; $1·4143^2 = 2·000\,244\,49$

Clearly $\sqrt{2}$ lies between 1·4142 and 1·4143

$1·4143^2 - 2 = ·000\,244\,49$; $2 - 1·4142^2 = ·000\,038\,36$

The next few figures for $\sqrt{2}$ can be obtained approximately by dividing
10 000 000 000 000 in the ratio 3836 to 24 449.
It is easier to use the ratio 38 to 244.

$$38 + 244 = 282 \qquad \frac{38}{282} = 1388888 \text{ (ignoring the decimal places)}$$

So an approximation to $\sqrt{2}$ will be 1·414 213 888.
By actual multiplication this will be found to be correct as far as the 3: the
8's are wrong.

Had we used the ratio 3836 to 24 449, we should have got $\dfrac{3836}{28\,285} = 1356196$.

So the next approximation to $\sqrt{2}$ is 1·41421356196.
By actual multiplication, this is correct as far as the first 6,
i.e. $\sqrt{2} = 1·414\,213\,56$.
If you had enough capacity you could apply proportional parts again and
get some more figures after the 6.

a) The tables give $\sqrt{3}$ as 1·732. Using the above method, see how many
more figures you can get for $\sqrt{3}$.

b) The tables give 2·236 for $\sqrt{5}$. Using the above method, see how many
more figures you can get for $\sqrt{5}$.

Note A more powerful method for finding $\sqrt{2}$, $\sqrt{3}$ etc. will be given in a
later book.

* 12 A proof that $\sqrt{2}$ is not rational
(See 15B, question 7)

If $\sqrt{2}$ is rational, let $\sqrt{2} = \dfrac{a}{b}$ where a and b are integers and the fraction $\dfrac{a}{b}$ has been cancelled down to its lowest possible terms, i.e. a and b contain no common factor.

Squaring gives $2 = \dfrac{a^2}{b^2}$

Therefore $2b^2 = a^2$.
Therefore a^2 is even. Hence a is even.
Let $a = 2k$, where k is an integer.

Then $\sqrt{2} = \dfrac{2k}{b}$. Squaring $2 = \dfrac{4k^2}{b^2}$ i.e. $2b^2 = 4k^2$, i.e. $b^2 = 2k^2$

Therefore b^2 is even. Therefore b is even.
Therefore both a and b are even. Therefore they have a factor 2 in common.

So starting with the assumption that $\sqrt{2} = \dfrac{a}{b}$ where a and b have no factors in common, we have proved they have a factor in common. So our original assumption that $\sqrt{2}$ could be represented by $\dfrac{a}{b}$ must be wrong, i.e. $\sqrt{2}$ cannot be represented by a rational fraction. It is in fact irrational.

   a) Can you give a similar proof that $\sqrt{3}$ is irrational?
   b) Can you give a similar proof that $\sqrt[3]{2}$ is irrational?

13 Blaise Pascal was born in France in 1625. By the time he was 12 years old he was wanting to know 'What is geometry?'. He discovered many theorems for himself. He is said to have folded a paper triangle in such a way that the angles met at a point on the base, and so found that the sum of the angles of a triangle was 180°.
Can you do this?

14 The cross number has vertical, horizontal and diagonal lines of symmetry.

Copy the grid on to squared paper and complete the shading. Number the 'starting' squares and then solve the puzzle using the given clues.

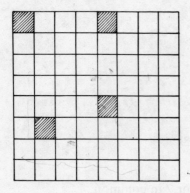

25 If $x \to \frac{2}{3}(x+1)^2$, what is 9 mapped on to?

26 Find the co-ordinates of the image of $(-4, 4)$, after a rotation of $90°$ clockwise about $(2, 2)$.

DOWN

1 $0.\dot{3}\dot{6} = \frac{4}{x}$. Find x.

3 $535 \div 6.34$ (to 2 s.f.)

4 $\sqrt{8649}$

5 If $3x - 4(x-7) = 3 + (x-1)$, find x.

6 $\dfrac{1.9 \times 10^8}{2 \times 10^6}$

8 $CE = 3$ cm
$ED = 5$ cm
$CB = 12$ cm
Find AB.

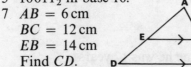

9 The 4th line of Pascal's triangle.

10 $\begin{pmatrix} 1 & 0 & 2 & 1 \\ 0 & 2 & 0 & 3 \\ 1 & 2 & 1 & 3 \\ 4 & 1 & 0 & 1 \end{pmatrix} \begin{pmatrix} 2 \\ 0 \\ 2 \\ 1 \end{pmatrix} = \begin{pmatrix} a \\ b \\ c \\ d \end{pmatrix} = \begin{pmatrix} \ \\ \ \\ \ \\ \ \end{pmatrix}$

Hint Multiply column by first row to get a. Multiply column by second row to get b. Multiply column by 3rd, 4th rows to get c, d.

11 Find the smallest share when £75 is divided in the ratio $8:4:3$.

13 The 10th term in the Fibonacci sequence.

16 An £x article is reduced by $12\frac{1}{2}\%$ in a sale to cost £28. Find x.

18 $\begin{pmatrix} 7 & 8 \\ 4 & 1 \end{pmatrix} - \begin{pmatrix} 3 & a \\ 1 & b \end{pmatrix} = \begin{pmatrix} 4 & 1 \\ 3 & 0 \end{pmatrix}$. Find $\begin{pmatrix} a \\ b \end{pmatrix}$

19 $x*y$ means $(2x+y)^2$. Find $5*(-2)$.

20 If $AC = 2$ m, find AB, in centimetres (to 2 s.f.)

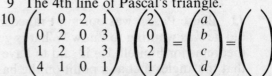

22 Change (in pence) from £5 after buying 27 litres of petrol at 18p a litre.

24 $(10^6 \times 10^3)^2 = 10^n$. Find n.

ACROSS

1 Co-ordinates of the image of $(-3, 2)$ after reflection in line $x = -1$.

2 $a*b$ means $3a^2 + b$. Find $2*6$.

4 35% of £2.60, in pence.

5 10011_2 in base 10.

7 $AB = 6$ cm
$BC = 12$ cm
$EB = 14$ cm
Find CD.

9 What is range of $x \to 2x + 3$ if the domain is $\{-1, 0, 1, 2\}$?

11 Simplify $(2x+9)+(x-1)-(3x-5)$.

12 A rectangle of area 8.75 cm^2 is enlarged so that the lengths are doubled. What is the new area?

14 $AB = 8$ cm, $BC = 6$ cm. Find $\angle ACB$ to the nearest degree.

15 The mean of $73, x, 76, 72$ is 74. Find x.

17 $1\frac{10}{13}$ as a decimal to 4 s.f. (ignore decimal point).

21 Largest integer in the set $\{x : (x-2)^2 < 400\}$

23 $AB = 50$ cm
$AC = 48$ cm
Find BC.

24 5.45×1.85, correct to 2 s.f.

146

15

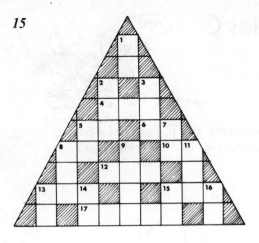

ACROSS

4 Sum of numbers in 8th row of Pascal's triangle.

5 Find x.

6 B is due east from A. The bearing of C from A is 028°, and from B is 306°. Find $\angle ACB$.

8 $AB = 6$ cm
$BC = 12$ cm
$BE = 8$ cm
Find CD

10 Difference between 5th and 7th triangle numbers.

12 Sum of angles in a triangle.

13 AC is a line of symmetry. $AC = BC$ and $\angle ABC = 39°$. Find reflex $\angle BCD$.

15 $\angle ABC = 44°$
$\angle EAC = 110°$
Find $\angle ACD$

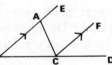

17 5th row of Pascal's triangle.

DOWN

1 In $\triangle ABC$, D is a point on BC such that AD bisects $\angle BAC$. $BD = AD$ and $\angle ADC = 80°$. Find $\angle ACD$.

2 $\triangle ABC$ is isosceles. The base CA is extended to P such that $\angle PBC = 90°$. If $\angle PBA = 50°$, find $\angle BAP$.

3 ABC is the cross section of a prism of length 10 cm. Find its total surface area.

5 $\cos 50° = \sin x$. Find x.

7 The diagonal of a square is 29·7 cm. Find the length of a side.

8 The volume of the prism in 3 down.

9 Give two consecutive triangle numbers with a sum of 64.

11 Find the length of AB in cm.

13 Product of order of symmetry and number of planes of symmetry of a prism whose cross section is an equilateral triangle.

14 In $\triangle ABC$, AB is 21·7 cm, BC is 25 cm, $\angle ABC$ is 90°. Find $\angle ACB$ to the nearest degree.

15 Find XZ to nearest whole number.

16 Find $\angle BAC$.

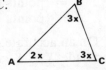

147

Miscellaneous Examples C

C1

1 Bars of chocolate are sold both individually and in packs of 4 bars. If a pack costs 25p, which is 1p less than the cost of 4 individual bars, find the cost of 11 bars, all bought singly.

2 B is a point on AC which divides it in the ratio $AB:BC = 1:2$. Calculate the height DB and hence find the area of triangle DAC. Find also $\angle DCB$ (to the nearest degree).

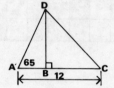

3 Two rings are attached to the back of a picture frame, 20 cm apart and 8 cm from the top of the frame. The picture is suspended by a loop of thread passing through the rings and over a hook on the wall. If the hook is to be 3 cm below the top of the frame, what is the length of thread in the loop?

4 Each of the following are either terminating or recurring decimals. Write each in decimal form, giving any group of recurring decimals in full, and hence arrange them in ascending order of size:

$$\frac{3}{4} \qquad \frac{2}{7} \qquad \frac{2}{3} \qquad \frac{4}{15} \qquad \frac{7}{9} \qquad \frac{1}{2} \qquad \frac{7}{8} \qquad \frac{9}{20}$$

5 A crate contains 500 oranges, 15 % of which were damaged in transit to the greengrocers.
If the oranges were originally to be sold at 6p each, how much less would the greengrocer receive if he sold the good ones at 7p each?

C2

1 a) Find 42 % of £3.
  b) On a menu there is a choice of two set meals, one costing £1.60 and the other £1.70. 8 % VAT is added to each. The final bill for a group of 4 people was £7.02. How many had the first set meal and how many the second?

2 $ABCD$ is a square of side 6 cm.

  $X = \{$points inside the square$\}$
  $Y = \{$points $P:P$ is nearer BC than $DC\}$
  $Z = \{$points $P:CP < 6\,$cm$\}$

On an accurate drawing, show the boundaries of X, Y and Z.
Shade $X \cap Y \cap Z$.

3 *a)* Divide £24 in the ratio 2:3:5.
b) In triangle *ACD*, *BE* is parallel
to *CD*. *AC* = 10 cm. *BE* = 3·6 cm.
B divides *AC* in the ratio *AB*:*BC*
= 3:2. Find the length of *CD*.

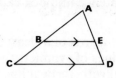

4 *AC* = 5 cm
BC = 3·5 cm
CD = 5·1 cm
$\angle ABC = \angle ACD = 90°$
Find $\angle ACB$ and show that *ABCD* is
a trapezium.

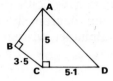

5 Write the following as fractions:

a) 0·75   *b)* 0·46   *c)* 0·4$\dot{6}$ *d)* 0·$\dot{1}\dot{2}$ *e)* 0·$\dot{8}$

C3

1 A jug holding 1 litre is 15 cm high. What is the height of a similar jug
which holds 125 ml?
Another jug, similar to the others, is 10 cm high. What fraction of a litre does
this hold?

2 From Stretton, Bretford is 5·1 km away on a bearing of 025°, and the
bearing of King's Newnham is 043°.
If King's Newnham is due east of Bretford, find how far east of Stretton is
King's Newnham and the direct distance between these two villages.

3 *D* is the midpoint of *AB*.
$\angle ABC = 90°$, *AC* = 8·3 cm,
BC = 4·6 cm.
Calculate the length of *DC*.

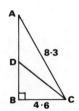

4 *a)* In an industrial dispute, union members are asking for pay rises of 25%.
If a man earns £3500 at present, what would he earn if a pay rise of 15%
is finally agreed?
b) The price of a carton of milk increases from 10p to 10½p. What
percentage increase is this?

5 Write a sentence in block capitals using only those letters which have line symmetry.

a) Find two words of at least 3 letters, in which all the letters have horizontal lines of symmetry.
b) Repeat *a)* with vertical lines of symmetry.
c) Repeat *a)* with rotational symmetry.
d) Can you find a word for which the whole word has point (half-turn rotational) symmetry about its mid-point?
e) Which letters have neither rotational nor line symmetry?
f) Draw a Venn diagram showing the grouping by symmetry of all the letters of the alphabet, written as block capitals.

C4

1 One angle of a right-angled triangle is 36°. If the length of the shortest side is 4·5 cm, find the length of the hypotenuse. If the length of the longer side is 4·5 cm, what is the length of the hypotenuse?

2 Say whether the following statements are true or false:

a) $\frac{4}{15} < \frac{2}{9}$ *b)* $121_5 > 212_4$
c) The point $(3, 5)$ lies on the line $y = 2x - 1$
d) There is just one number between 10 and 50 which is both a square number and a triangle number.
e) To convert millimetres to metres, multiply by 1000.

3 Last autumn, I bought a refrigerator which cost £59.50 including VAT at $12\frac{1}{2}$%. Find, to the nearest penny, the amount of tax paid.

4 *FBDE* and *GCDE* are both parallelograms and *G* is the mid-point of *FB*.
Given than $\vec{ED} = 2a$ and $\vec{DC} = b$, write the following in terms of *a* and *b*

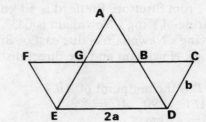

a) $\vec{BC}$ *d)* $\vec{EA}$

b) $\vec{FB}$ *e)* $\vec{DA}$

c) $\vec{FC}$ *f)* $\vec{BA}$

5 *a)* If $a = 1·555\ldots$ find $10a$.
By subtraction find the value of $9a$ and hence '*a*' expressed as a fraction in its simplest form.
b) If $b = 2·5454\ldots$ find $100b$.
By subtraction find the value of $99b$ and hence '*b*' expressed as a fraction in its simplest form.
c) Use this method to change the following to fractions:
$2·1666\ldots, \quad 1·2444\ldots, \quad 1·123\ 123\ldots$

C5

1 Calculate the following:

a) 5% of £2.40

b) The marks a candidate must score out of 120 in order to pass an examination in which 45% is the pass mark.

c) The total number of girls in a school if there are 75 in the sixth form and this is 12% of the total.

2 A ladder is 3·25 metres long, and stands on horizontal ground. How far from a wall must its foot be placed for the top of the ladder to reach a window 3 metres up the wall?

If the foot of the ladder is 1·5 metres from the wall, how far short of the window is the top of the ladder?

3 A model of some new buildings is made on the scale of 5 cm to 1 metre.

a) How long is the model if the building is 50 metres long?

b) The area of a lawn in front of the model is 16 000 cm². How much lawn will there be in front of the building?

c) Write the scale of the model in the form 1 : *n*.

d) Write the comparison between the actual volume and the volume of the model in ratio form.

4 Use your slide rule to find answers to the following questions. Give your answers to 2 s.f. If you feel sure of a third figure, give it in brackets.

a) What is 17 as a percentage of 27?

b) What is $\frac{7}{19}$ as a decimal?

c) What is 39% of £13.40?

d) What number multiplied by itself gives 57·6?

e) By what must you multiply 17·3 to get an answer equal to the product of 51·9 and 24·6?

5 A triangle has vertices *A* (2, 2), *B* (3, 5) and *C* (5, 2). It is translated by the vector $\binom{3}{2}$ to *A′B′C′* and then rotated through 90° anticlockwise about (2, 2) to give *A″B″C″*. State the co-ordinates of *A′, B′ C′, A″, B″* and *C″*.

State also the centre of a rotation that will map *ABC* directly on to *A″B″C″* without translation. What is the angle of this rotation?